Strong Interactions in Spacelike and Timelike Domains

Strong Interactions in Spacelike and Timelike Domains

Dispersive Approach

Alexander V. Nesterenko
Bogoliubov Laboratory of Theoretical Physics
Joint Institute for Nuclear Research
Dubna, Moscow region, Russian Federation

AMSTERDAM • BOSTON • HEIDELBERG • LONDON • NEW YORK • OXFORD
PARIS • SAN DIEGO • SAN FRANCISCO • SINGAPORE • SYDNEY • TOKYO
ELSEVIER

Elsevier
Radarweg 29, PO Box 211, 1000 AE Amsterdam, Netherlands
The Boulevard, Langford Lane, Kidlington, Oxford OX5 1GB, United Kingdom
50 Hampshire Street, 5th Floor, Cambridge, MA 02139, United States

Notices
Knowledge and best practice in this field are constantly changing. As new research and experience broaden our understanding, changes in research methods, professional practices, or medical treatment may become necessary.

Practitioners and researchers must always rely on their own experience and knowledge in evaluating and using any information, methods, compounds, or experiments described herein. In using such information or methods they should be mindful of their own safety and the safety of others, including parties for whom they have a professional responsibility.

To the fullest extent of the law, neither the Publisher nor the authors, contributors, or editors, assume any liability for any injury and/or damage to persons or property as a matter of products liability, negligence or otherwise, or from any use or operation of any methods, products, instructions, or ideas contained in the material herein.

Library of Congress Cataloging-in-Publication Data
A catalog record for this book is available from the Library of Congress

British Library Cataloguing-in-Publication Data
A catalogue record for this book is available from the British Library

ISBN: 978-0-12-803439-2

For information on all Elsevier publications
visit our website at https://www.elsevier.com/

Working together
to grow libraries in
developing countries

www.elsevier.com • www.bookaid.org

Publisher: John Fedor
Acquisition Editor: Anita Koch
Editorial Project Manager: Amy Clark
Production Project Manager: Paul Prasad Chandramohan
Cover Designer: Mathew Limbert

Typeset by SPi Global, India

Contents

6. *R*-Ratio at High Energies

About the Author

Dr. Alexander V. Nesterenko is a senior researcher at the Bogoliubov Laboratory of Theoretical Physics, Joint Institute for Nuclear Research, Dubna, Russian Federation. He graduated with honors from Moscow State University where he obtained a PhD in Theoretical Physics. He was a postdoctoral researcher at the École Polytechnique, France, and University of Valencia, Spain. He actively works in the area of the theoretical particle physics and is a referee for several APS and IOP journals. An experienced lecturer in Quantum Field Theory and Quantum Chromodynamics, he has published two textbooks based on his lecture course.

Preface

The landmark discovery of the asymptotic freedom in Quantum Chromodynamics (QCD) has eventually resulted in an overwhelming success of the perturbative description of the high-energy strong interaction processes in the spacelike (Euclidean) domain. At the same time the QCD perturbation theory appears to be directly inapplicable to the studies of hadron dynamics in the timelike (Minkowskian) domain. Furthermore, the low-energy strong interaction processes entirely fall out of the scope of perturbative approach and can be handled only within nonperturbative methods.

In general, there is a diversity of the nonperturbative approaches to QCD, which originate in various aspects of the tangled dynamics of the colored fields. A certain clue on the low-energy hadron dynamics is supplied by the corresponding dispersion relations, which convert the underlying kinematic restrictions on a physical process on hand into the intrinsically nonperturbative constraints on the related functions that should definitely be accounted for when one goes beyond the scope of perturbation theory. Furthermore, the dispersion relations provide the only way of the self-consistent description of the strong interaction processes in the timelike domain.

The book primarily focuses on the mutually interrelated hadronic vacuum polarization function, R-ratio of electron-positron annihilation into hadrons, and the Adler function, which govern various strong interaction processes in the spacelike and timelike domains at the energy scales spanning from low to high energies. Specifically, the book presents the essentials of the dispersion relations for these functions, recaps their perturbative calculation, and delineates the dispersively improved perturbation theory. The latter merges the nonperturbative constraints imposed by the pertinent dispersion relations on the functions on hand with corresponding perturbative input in a self-consistent way, gets rid of innate obstacles of the QCD perturbation theory, and considerably extends its applicability range toward the low energies. The book also elucidates the basics of the continuation of the spacelike perturbative results into the timelike domain, which is essential for the study of electron-positron annihilation into hadrons and the relevant strong interaction processes.

The book is based on a portion of a course of lectures on QCD, which was delivered by the author at the Moscow State University and at the Moscow Institute of Physics and Technology. The topics covered in the book, being of a direct relevance to the numerous ongoing research programs in particle physics and future collider projects, can be of benefit to graduate and postgraduate students, academic lecturers, and scientific researchers working in the related fields of theoretical and mathematical physics.

A.V. Nesterenko
Dubna, Russia
June 2016

Acknowledgments

The author expresses his heartfelt gratitude to A.C. Aguilar, A.B. Arbuzov, B.A. Arbuzov, A.P. Bakulev, G.S. Bali, V.V. Belokurov, A.V. Borisov, N. Brambilla, G. Cvetic, F. De Fazio, A.E. Dorokhov, H.M. Fried, R. Kaminski, A.L. Kataev, D.I. Kazakov, S.A. Larin, M. Loewe, K. Milton, S. Narison, J. Papavassiliou, M. Passera, J. Portoles, G.M. Prosperi, F. Schrempp, A.V. Sidorov, C. Simolo, I.L. Solovtsov, O.P. Solovtsova, O.V. Teryaev, A. Vairo, and H. Wittig for the interest, stimulating comments, valuable advices, fruitful discussions, and continuous support.

Introduction

The theoretical description of the strong interaction processes is utterly based on the quantum non-Abelian gauge field theory, namely, Quantum Chromodynamics (QCD). This theory originates in the notion of quarks and gluons. The former are the constituents of all hadrons that were proposed in the mid-1960s in the milestone works by Gell-Mann [1], Zweig [2, 3], and Petermann [4] (see also paper [5]), whereas the latter are the quanta of a massless gauge vector field providing interaction between quarks. Both quarks and gluons carry a specific quantum number, which plays the crucial role in hadron physics. This quantum number was also suggested in the mid-1960s in the papers by Greenberg [6], Han and Nambu [7, 8], Bogoliubov, Struminsky, and Tavkhelidze [9, 10], and named "color" afterward [11] (see also papers [12–14]).

Thorough theoretical and experimental investigations revealed that the strong interactions possess two distinctive features. First, the strength of the interaction between colored objects decreases when the characteristic energy of the process on hand increases. In other words, the QCD invariant charge $\alpha_s = g^2/(4\pi)$, which is also called the strong coupling, vanishes at large momenta transferred, which constitutes the asymptotic freedom of QCD. Second, free quarks and gluons, as well as any other colored final state, have never been observed experimentally, which constitutes the so-called confinement of color.

These two phenomena are governed by the strong interactions at the opposite energy scales. Specifically, the asymptotic freedom takes place at high energies, or in the so-called ultraviolet domain, that corresponds to small spatial quark separations. As for the confinement of color it is related to low energies, or the so-called infrared domain, that, in turn, corresponds to large spatial quark separations.

The asymptotic freedom in the non-Abelian gauge field theory was discovered in the early 1970s by 't Hooft [15], Gross and Wilczek [16], and Politzer [17] (see also papers [18–21]). In turn this finding gave rise to an extensive employment of perturbation theory in the study of the strong interaction processes at high energies. However, in practice the results of perturbative calculations in QCD appear to be of a limited applicability. This is primarily caused by the fact that the perturbative approach in Quantum Field Theory entirely relies on the assumption that the value of the corresponding invariant charge is small enough. In particular, it is this assumption that allows one to approximate an experimentally measurable physical observable by perturbative power series in the respective coupling, which drastically simplifies the theoretical analysis of the process on hand. In Quantum Electrodynamics (QED) the aforementioned assumption is valid for all experimentally accessible energies[1] that makes the QED perturbative calculations reliable for all practical purposes. On the contrary, in QCD the foregoing assumption of smallness of the perturbative strong running coupling is valid for high and intermediate energies only. In the experimentally accessible infrared domain the perturbative QCD comes out of the "small coupling" regime, which makes perturbation theory inapplicable to the study of the strong interaction processes at low energies. In other words, the theoretical description of hadron dynamics in the infrared domain wholly remains beyond the scope of perturbation theory and can only be performed by making use of nonperturbative methods. Additionally, it is necessary to outline that the QCD perturbation theory is directly applicable to the study of the strong interaction processes only in the spacelike (Euclidean) domain, whereas the self-consistent description of hadron dynamics in the timelike (Minkowskian) domain is based on pertinent dispersion relations.

In fact, the theoretical description of the strong interaction processes at low energies remains one of the most challenging issues of the elementary particle physics since the discovery of the QCD asymptotic freedom. In general, there is a variety of methods that enable one to explore the tangled dynamics of colored fields in the infrared domain and to effectively describe the nonperturbative aspects of the strong interactions. In particular, over the past decades the confinement of color has been addressed in the framework of such approaches as, for example, the string models of hadrons [22–26], the phenomenological potential models of quark-antiquark interaction [27–32], the quark bag models [33–35], the analytic gauge-invariant QCD [36–41], the holographic QCD [42–45], the lattice simulations [46–54], and many others.

Theoretical particle physics widely employs various methods originated in the corresponding dispersion relations (see, e.g., books [55–59] and papers [60–63]). In particular, these relations have proved their efficiency in such issues as the extension of the applicability range of chiral perturbation theory [64–74], the precise determination of parameters of resonances [75, 76], the assessment of the hadronic light-by-light scattering [77–80], as well as many others. Basically, the dispersion relations provide an additional source of the information on the pertinent physical processes, which does not rely on the perturbation theory. Specifically, such relations convert the physical kinematic restrictions on the process on hand into the intrinsically nonperturbative constraints on the related functions that should definitely be accounted for when one comes out of the applicability range of the perturbative approach. Furthermore, as noted earlier, the relevant dispersion relations provide the only way to properly analyze the strong interaction processes in the timelike domain.

---

1. In QED the perturbative approach fails in the ultraviolet asymptotic at the energy scale of the order of the Planck mass (see also Section 2.2).

A key role in the study of various strong interaction processes in the spacelike and timelike domains is played by the so-called hadronic vacuum polarization function $\Pi(q^2)$, the related function $R(s)$, and the Adler function $D(Q^2)$. In particular, these functions govern, for example, such processes as the electron-positron annihilation into hadrons, inclusive τ lepton and Z boson hadronic decays, as well as the hadronic contributions to such observables of the precision particle physics as the muon anomalous magnetic moment $(g-2)_\mu$ and the running of the electromagnetic fine structure constant. The theoretical analysis of these processes constitutes a decisive self-consistency test of QCD and the entire Standard Model, that, in turn, puts strong limits on a possible new fundamental physics beyond the latter. Additionally, the aforementioned strong interaction processes are characterized by the energy scales spanning from low to high energies, so that their theoretical exploration constitutes a natural framework for a thorough investigation of both perturbative and intrinsically nonperturbative aspects of hadron dynamics. It is also worthwhile to note that most of the processes mentioned earlier are of direct relevance to the physics at the future collider projects, such as the Future Circular Collider (FCC) at CERN (specifically, its FCC-ee part), the Circular Electron-Positron Collider (CEPC) in China (specifically, the first phase of this project), the International Linear Collider (ILC), the Compact Linear Collider (CLIC), as well as the E989 experiment at Fermilab, the E34 experiment at Japan Proton Accelerator Research Complex (J-PARC), and others.

The book mainly focuses on the description of the aforementioned hadronic vacuum polarization function $\Pi(q^2)$, the function $R(s)$, which is also called R-ratio of electron-positron annihilation into hadrons, and the Adler function $D(Q^2)$. In particular, the book highlights the basics of the dispersion relations for these functions, delineates their perturbative calculation, and presents the dispersively improved perturbation theory (DPT). The latter merges the intrinsically nonperturbative constraints for the functions on hand, which originate in the pertinent dispersion relations, with corresponding perturbative input in a self-consistent way. In addition, it overcomes inherent obstacles of the QCD perturbation theory and substantially extends its applicability range toward the infrared domain. The book also elucidates the essentials of the continuation of the spacelike perturbative results into the timelike domain, which is indispensable for the study of the electron-positron annihilation into hadrons and the related strong interaction processes.

The layout of the book is as follows. Chapter 1 presents the basics of the dispersion relations for the functions on hand. Specifically, Section 1.1 describes the function $R(s)$, derives in the leading order of perturbation theory the total cross sections of the electron-positron annihilation into a muon-antimuon pair as well as into hadrons, discusses the kinematic restrictions for the process on hand, and depicts the experimental data on R-ratio. Section 1.2 deals with the hadronic vacuum polarization function $\Pi(q^2)$, discusses its properties in the complex q^2-plane, derives the dispersion relation for $\Pi(q^2)$, and presents the experimental prediction for the latter. Section 1.3 recounts the Adler function $D(Q^2)$, derives the corresponding dispersion relations, discusses the nonperturbative constraints imposed by the latter, and portrays the experimental prediction for $D(Q^2)$. Chapter 2 delineates the perturbative strong running coupling and its basic features. In particular, Section 2.1 describes the renormalization group equation for the QCD invariant charge, discusses the perturbative calculation of the corresponding β function up to the five-loop level, and briefly recaps the estimations of its coefficients, which have not been explicitly calculated yet. Section 2.2 derives the solution of the renormalization group equation for the strong running coupling at the one-loop level, examines its properties, and describes the so-called matching procedure. Section 2.3 presents the solutions of the renormalization group equation for the QCD invariant charge at the higher loop levels and discusses their peculiarities. Chapter 3 elucidates the calculation of the functions on hand within perturbative approach. Specifically, Section 3.1 derives the hadronic vacuum polarization function $\Pi(q^2)$ in the leading order in the strong coupling, discusses the corresponding perturbative corrections up to the four-loop level, and recounts the general features of the obtained expressions for $\Pi(q^2)$. Section 3.2 presents the perturbative approximation for the Adler function $D(Q^2)$ at the first four loop levels, describes its basic peculiarities, and briefly discusses the estimations of the uncalculated yet higher-order corrections. Section 3.3 derives the R-ratio of electron-positron annihilation into hadrons in the leading order of perturbation theory, concisely recaps the most salient features of the perturbative approximation of the function $R(s)$ at the higher loop levels, and expounds its general peculiarities. Chapter 4 presents the essentials of the dispersive approach to QCD. Specifically, Section 4.1 derives the unified integral representations for $\Pi(q^2)$, $R(s)$, and $D(Q^2)$, which embody all the nonperturbative constraints imposed by the pertinent dispersion relations and express the functions on hand in terms of the common spectral density. Section 4.2 provides the explicit expression for the perturbative spectral density at an arbitrary loop level and describes its basic features. Section 4.3 discusses the obtained integral representations for the functions on hand in the massless limit and recounts its implications. Chapter 5 delineates the calculation of the functions $\Pi(q^2)$, $R(s)$, and $D(Q^2)$ within dispersively improved perturbation theory and elucidates their salient distinctions. Specifically, Section 5.1 studies the hadronic vacuum polarization function $\Pi(q^2)$ at the one-loop level, discusses the pertinent higher loop corrections, compares the function $\Pi(q^2)$ with its experimental prediction, and recounts the scheme stability of the obtained results. Section 5.2 expounds the function $R(s)$ at the one-loop level, studies the higher loop corrections to $R(s)$, examines its scheme stability, and juxtaposes the obtained results with experimental data on R-ratio by making use of the smearing method. Section 5.3 recounts the Adler function $D(Q^2)$ at the one-loop level, examines the relevant higher loop corrections, compares the function $D(Q^2)$ with its experimental prediction, and expounds the scheme stability of the obtained results. Chapter 6 elucidates the perturbative approximation of the R-ratio of electron-positron annihilation into hadrons and discusses its basic peculiarities. Specifically, Section 6.1 studies the reexpansion of the function $R(s)$ at high energies and describes the appearance of the π^2-terms. Section 6.2 delineates the perturbative expression for the R-ratio beyond the three-loop level, recounts its convergence range, and discusses the impact of the π^2-terms. Section 6.3 expounds the estimation of the uncalculated yet coefficients of the perturbative approximation of R-ratio. The technical auxiliary materials are gathered in the Appendices. Specifically, Appendix A supplies the explicit expressions for the perturbative strong running coupling at the higher loop levels. Appendix B presents the explicit expressions for the aforementioned spectral function at the higher orders of perturbation theory and discusses the numerical evaluation of the functions $\Pi(q^2)$, $R(s)$, and $D(Q^2)$ within DPT. Appendix C recaps the perturbative approximation of the R-ratio of electron-positron annihilation into hadrons and provides the explicit expressions for the π^2-terms at the higher-loop levels.

Chapter 1

Basic Dispersion Relations

The hadronic vacuum polarization function, as well as the related Adler function and the R-ratio of electron-positron annihilation into hadrons, plays a significant role in various issues of contemporary elementary particle physics. The theoretical description of these functions is inherently based on the perturbation theory, which supplies the corresponding high-energy input, and dispersion relations, which embody the pertinent low-energy kinematic constraints and enable one to study the hadron dynamics in the timelike domain in a self-consistent way. The functions on hand form the basis of the theoretical analysis of a variety of the strong interaction processes, including the hadronic contributions to precise electroweak observables, that, in turn, constitutes a decisive test of Quantum Chromodynamics (QCD) and entire Standard Model and imposes strict restrictions on a possible new fundamental physics beyond the latter.

1.1 *R*-RATIO OF ELECTRON-POSITRON ANNIHILATION INTO HADRONS

The process of electron-positron annihilation into hadrons plays a distinctive role in elementary particle physics. This is primarily caused by a remarkable fact that its theoretical description requires no phenomenological models of the process of hadronization, which forms the experimentally detected final-state particles. In turn, this feature makes it possible to extract the key parameters of the theory from pertinent data in the model-independent way, that is certainly essential for the experimental verification of the latter.

The analysis of the process on hand usually involves the so-called R-ratio of electron-positron annihilation into hadrons, which is defined as the ratio[1] of two experimentally measurable total cross sections

$$R(s) = \frac{\sigma\left(e^+e^- \to \text{hadrons}; s\right)}{\sigma\left(e^+e^- \to \mu^+\mu^-; s\right)}. \tag{1.1}$$

---

1. It is assumed that both total cross sections on the right-hand side of Eq. (1.1) correspond to the leading order in the electromagnetic fine structure constant $\alpha_{\text{em}} = e^2/(4\pi)$.

Strong Interactions in Spacelike and Timelike Domains. http://dx.doi.org/10.1016/B978-0-12-803439-2.00001-6

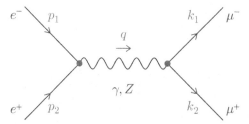

FIG. 1.1 The process of the electron-positron annihilation into muon-antimuon pair in the lowest order of perturbation theory.

Such normalization proves to be particularly convenient for the studies of the hadronic vacuum polarization. Specifically, the leptonic parts of the processes contributing to Eq. (1.1) are identical to each other that makes the R-ratio strictly focused on the effects due to the strong interactions.

Let us first address the process of the electron-positron annihilation into a muon-antimuon pair (see Fig. 1.1). The matrix element corresponding to the diagram with an intermediate photon takes the form (see also papers by Bhabha [81] and Feynman [82, 83]):

$$M_{\text{fi}} = i\bar{v}^{[e]}(p_2, \sigma_2)\gamma^\mu u^{[e]}(p_1, \sigma_1)\frac{e^2}{q^2}\left[g_{\mu\nu} - (1-\xi)\frac{q_\mu q_\nu}{q^2}\right]\bar{u}^{[\mu]}(k_1, \lambda_1)\gamma^\nu v^{[\mu]}(k_2, \lambda_2)\text{.} \tag{1.2}$$

In this equation e is the charge of the electron, $g_{\mu\nu} = \text{diag}(1, -1, -1, -1)$ stands for the metric tensor, γ_μ denotes the Dirac matrix, $q = p_1 + p_2$ is the timelike kinematic variable, and ξ stands for the gauge parameter (a real constant), whereas $u^{[e]}(p_1, \sigma_1)$, $\bar{v}^{[e]}(p_2, \sigma_2)$, $\bar{u}^{[\mu]}(k_1, \lambda_1)$, and $v^{[\mu]}(k_2, \lambda_2)$ denote the Dirac spinors of incoming electron with four-momentum p_1 and spin σ_1, the incoming positron with four-momentum p_2 and spin σ_2, the outgoing muon with four-momentum k_1 and spin λ_1, and the outgoing antimuon with four-momentum k_2 and spin λ_2, respectively. In what follows the Feynman gauge ($\xi = 1$) will be assumed, which makes the expression for the photon propagator quite simple. The cross section of the process on hand is proportional to the matrix element (1.2) squared, which can be represented in the following way:

$$|M_{\text{fi}}|^2 = \frac{e^4}{s^2}\bar{u}^{[e]}(p_1, \sigma_1)\gamma_\nu v^{[e]}(p_2, \sigma_2)\bar{v}^{[e]}(p_2, \sigma_2)\gamma_\mu u^{[e]}(p_1, \sigma_1)$$
$$\times \bar{v}^{[\mu]}(k_2, \lambda_2)\gamma^\nu u^{[\mu]}(k_1, \lambda_1)\bar{u}^{[\mu]}(k_1, \lambda_1)\gamma^\mu v^{[\mu]}(k_2, \lambda_2), \tag{1.3}$$

where $s = q^2 > 0$. Assuming that the spins of the final state particles are unknown and that the colliding beams are not polarized, one has to sum Eq. (1.3) over the spins of muon and antimuon and average the result over the spins of electron and positron. This eventually yields

$$\frac{1}{4} \sum_{\text{spin}} |M_{\text{fi}}|^2 = \frac{e^4}{4s^2} \, \text{Tr} \left[(\hat{p}_1 + m_e) \gamma_\nu (\hat{p}_2 - m_e) \gamma_\mu \right]$$

$$\times \, \text{Tr} \left[(\hat{k}_2 - m_\mu) \gamma^\nu (\hat{k}_1 + m_\mu) \gamma^\mu \right], \tag{1.4}$$

with

$$\sum_\sigma u_\alpha(p,\sigma) \bar{u}^\beta(p,\sigma) = (\hat{p} + m)^\beta_\alpha, \quad \sum_\sigma v_\alpha(p,\sigma) \bar{v}^\beta(p,\sigma) = (\hat{p} - m)^\beta_\alpha \tag{1.5}$$

being employed. In Eq. (1.4) $m_e \simeq 0.511 \, \text{MeV}$ and $m_\mu \simeq 105.66 \, \text{MeV}$ stand for the masses of electrons and muons, respectively [84]. Then by making use of the relations

$$\text{Tr} \left(\gamma_\mu \gamma_\nu \right) = 4 g_{\mu\nu}, \tag{1.6}$$

$$\text{Tr} \prod_{j=1}^{2r+1} \gamma_{\mu_j} = 0, \quad r = 0,1,2,3,\ldots, \tag{1.7}$$

$$\text{Tr} \left(\gamma_\mu \gamma_\nu \gamma_\rho \gamma_\sigma \right) = 4 \left(g_{\mu\nu} g_{\rho\sigma} - g_{\mu\rho} g_{\nu\sigma} + g_{\mu\sigma} g_{\nu\rho} \right), \tag{1.8}$$

one arrives at

$$\text{Tr} \left[(\hat{p}_1 + m_e) \gamma_\nu (\hat{p}_2 - m_e) \gamma_\mu \right] = 4 \left[p_{1\mu} p_{2\nu} + p_{2\mu} p_{1\nu} - (p_1 p_2) g_{\mu\nu} - m_e^2 g_{\mu\nu} \right] \tag{1.9}$$

and

$$\text{Tr} \left[(\hat{k}_2 - m_\mu) \gamma^\nu (\hat{k}_1 + m_\mu) \gamma^\mu \right] = 4 \left[k_1^\mu k_2^\nu + k_2^\mu k_1^\nu - (k_1 k_2) g^{\mu\nu} - m_\mu^2 g^{\mu\nu} \right]. \tag{1.10}$$

Thus the expression (1.4) acquires the following form:

$$\frac{1}{4} \sum_{\text{spin}} |M_{\text{fi}}|^2 = \frac{8e^4}{s^2} \left[(p_1 k_1)(p_2 k_2) + (p_1 k_2)(p_2 k_1) + m_e^2 (k_1 k_2) \right.$$

$$\left. + m_\mu^2 (p_1 p_2) + 2 m_e^2 m_\mu^2 \right]. \tag{1.11}$$

In what follows, it is convenient to study the process on hand in the center-of-mass reference frame. In this case the four-momenta of the involved particles read

$$p_1 = (E, \vec{p}), \quad p_2 = (E, -\vec{p}), \quad k_1 = (E, \vec{k}), \quad k_2 = (E, -\vec{k}). \tag{1.12}$$

In turn, the center-of-mass energy squared s is

$$s = (p_1 + p_2)^2 = q^2 = (k_1 + k_2)^2 = 4E^2, \tag{1.13}$$

the pertinent three-momenta are

$$\vec{p}^2 = \frac{s}{4} - m_e^2, \quad \vec{k}^2 = \frac{s}{4} - m_\mu^2, \tag{1.14}$$

and

$$(p_1 k_1) = (p_2 k_2) = \frac{s}{4} \left(1 - \cos\varphi \sqrt{1 - \frac{4m_e^2}{s}} \sqrt{1 - \frac{4m_\mu^2}{s}} \right), \tag{1.15}$$

$$(p_1 k_2) = (p_2 k_1) = \frac{s}{4} \left(1 + \cos\varphi \sqrt{1 - \frac{4m_e^2}{s}} \sqrt{1 - \frac{4m_\mu^2}{s}} \right), \tag{1.16}$$

$$(p_1 p_2) = \frac{s}{2} - m_e^2, \quad (k_1 k_2) = \frac{s}{2} - m_\mu^2, \tag{1.17}$$

where φ denotes the angle between the three-momentum of electron $\vec{p}$ and the three-momentum of muon $\vec{k}$. Hence, Eq. (1.11) can be represented in the following way:

$$\frac{1}{4} \sum_{\text{spin}} |M_{\text{fi}}|^2 = e^4 \left[1 + \cos^2\varphi \left(1 - \frac{4m_e^2}{s} \right) \left(1 - \frac{4m_\mu^2}{s} \right) + \frac{4}{s} \left(m_e^2 + m_\mu^2 \right) \right]. \tag{1.18}$$

Thus in the center-of-mass reference frame the differential cross section of the process on hand can be represented as

$$\frac{d\sigma \left(e^+ e^- \to \mu^+ \mu^-; s \right)}{d\Omega} = \frac{1}{(4\pi)^2} \frac{1}{s} \frac{|\vec{k}|}{|\vec{p}|} \frac{1}{4} \sum_{\text{spin}} |M_{\text{fi}}|^2 = \frac{e^4}{(4\pi)^2} \frac{1}{4s} \left(\frac{s - 4m_\mu^2}{s - 4m_e^2} \right)^{1/2}$$

$$\times \left[1 + \cos^2\varphi \left(1 - \frac{4m_e^2}{s} \right) \left(1 - \frac{4m_\mu^2}{s} \right) + \frac{4}{s} \left(m_e^2 + m_\mu^2 \right) \right]. \tag{1.19}$$

In the limit of a massless electron ($m_e = 0$) and muon ($m_\mu = 0$) this equation acquires the form

$$\frac{d\sigma \left(e^+ e^- \to \mu^+ \mu^-; s \right)}{d\Omega} = \frac{\alpha_{\text{em}}^2}{4s} \left(1 + \cos^2\varphi \right), \tag{1.20}$$

where $\alpha_{\text{em}} = e^2/(4\pi)$ stands for the electromagnetic fine structure constant. Therefore, the total cross section of the electron-positron annihilation into muon-antimuon pair reads

$$\sigma \left(e^+ e^- \to \mu^+ \mu^-; s \right) = \frac{\alpha_{\text{em}}^2}{4s} \int_0^{2\pi} d\phi \int_{-1}^1 \left(1 + \cos^2\varphi \right) d\cos\varphi = \frac{4\pi}{3} \frac{\alpha_{\text{em}}^2}{s}. \tag{1.21}$$

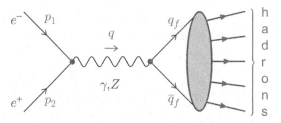

FIG. 1.2   The process of the electron-positron annihilation into hadrons in the lowest order of perturbation theory.

The obtained expression for the cross section of the process on hand (1.21) is a rapidly decreasing function of the center-of-mass energy squared s. The electromagnetic running coupling α_{em} entering Eq. (1.21), as well as its dependence on the energy scale, was thoroughly studied in the works by Burkhardt and Pietrzyk [85–89], Jegerlehner [90–93], and Hagiwara et al. [94–97], see also review [98] and references therein. Note that the process of electron-positron annihilation into muon-antimuon pair also involves intermediate Z boson ($M_Z \simeq 91.188\,\text{GeV}$ [84]), which has to be accounted for separately (see Fig. 1.1).

Let us turn now to the process of the electron-positron annihilation into hadrons. In the leading order of perturbation theory the latter is governed by the diagram displayed in Fig. 1.2. The matrix element corresponding to the diagram with intermediate photon takes the form

$$M_{\text{fi}} = i\bar{v}^{[e]}(p_2, \sigma_2)e\gamma^{\mu}u^{[e]}(p_1, \sigma_1)\frac{g_{\mu\nu}}{q^2}\langle\Gamma|J^{\nu}(q)|0\rangle. \tag{1.22}$$

As before, in this equation $u^{[e]}(p_1, \sigma_1)$ and $\bar{v}^{[e]}(p_2, \sigma_2)$ stand for the Dirac spinors of incoming electron with four-momentum p_1 and spin σ_1 and incoming positron with four-momentum p_2 and spin σ_2, respectively, $J_{\nu} = \sum_{f=1}^{n_f} Q_f{:}\bar{q}_f\gamma_{\nu}q_f{:}$ denotes the electromagnetic quark current, n_f is the number of active flavors,[2] Q_f stands for the electric charge of the quark of fth flavor (in units of the elementary charge e), and Γ is a hadronic final state. The photon propagator in Eq. (1.22) corresponds to the Feynman gauge, whereas $q^2 = s = (p_1 + p_2)^2 > 0$ denotes the respective timelike kinematic variable, namely, the center-of-mass energy squared.

Following the very same lines as earlier, it is quite straightforward to demonstrate that the total cross section of electron-positron annihilation into hadrons can be represented as

$$\sigma\left(e^+e^- \to \text{hadrons}; s\right) = 8\pi^2\frac{\alpha_{em}^2}{s^3}L_{\mu\nu}H_{\mu\nu}. \tag{1.23}$$

---

2. The number of active flavors is the number of quark flavors whose masses are less than the characteristic energy of a process on hand, see Table 1.1 and Section 2.1.

TABLE 1.1 The Electric Charges (in Units of the Elementary Charge) and Masses of Quarks [84]

f	Q_f	m_f
u	2/3	$2.3^{+0.7}_{-0.5}$ MeV[a]
d	−1/3	$4.8^{+0.5}_{-0.3}$ MeV[a]
s	−1/3	95 ± 5 MeV[a]
c	2/3	1.275 ± 0.025 GeV[b]
b	−1/3	4.18 ± 0.03 GeV[b]
t	2/3	173.21 ± 0.87 GeV[c]

[a] Current quark mass in a mass-independent subtraction scheme at $\mu = 2$ GeV.
[b] Running mass in the $\overline{MS}$ scheme.
[c] Direct measurement.

In this equation $L_{\mu\nu}$ denotes the leptonic tensor (the electron is assumed to be massless hereinafter)

$$L_{\mu\nu} = \frac{1}{4} \sum_{\text{spin}} \bar{v}^{[e]}(p_2, \sigma_2) \gamma_\mu u^{[e]}(p_1, \sigma_1) \left[\bar{v}^{[e]}(p_2, \sigma_2) \gamma_\nu u^{[e]}(p_1, \sigma_1) \right]^*$$

$$= \frac{1}{2} \left[q_\mu q_\nu - g_{\mu\nu} q^2 - (p_1 - p_2)_\mu (p_1 - p_2)_\nu \right], \tag{1.24}$$

and $H_{\mu\nu}$ stands for the hadronic tensor

$$H_{\mu\nu} = (2\pi)^4 \sum_\Gamma \delta(p_1 + p_2 - p_\Gamma) \langle 0|J_\mu(-q)|\Gamma\rangle \langle \Gamma|J_\nu(q)|0\rangle$$

$$= \int d^4 x\, e^{iqx} \langle 0|[J_\mu(x), J_\nu(0)]|0\rangle, \quad q^2 \geq 4m_\pi^2, \tag{1.25}$$

where p_Γ denotes the four-momentum of a final hadronic state Γ and $m_\pi \simeq 139.57$ MeV stands for the mass of π meson [84]. Eq. (1.25) employs the completeness of set of hadronic states

$$\sum_\Gamma |\Gamma\rangle \langle \Gamma| = \theta(q^2 - 4m_\pi^2), \tag{1.26}$$

with $\theta(x)$ being the Heaviside unit step function:

$$\theta(x) = \begin{cases} 1, & x \geq 0, \\ 0, & x < 0 \end{cases} \tag{1.27}$$

(note that $\theta(x)$ is a dimensionless function, though its argument may be of the dimension of, e.g., mass squared). Basically, Eq. (1.26) explicitly expresses the physical fact that the hadrons can be produced in the process of electron-positron annihilation only if the respective energy $\sqrt{s}$ is high enough, otherwise the matrix element of the process on hand (1.22) and its square vanish (see also discussion of this issue in, e.g., book [99] and papers [100, 101]). It is also worthwhile to mention that the processes of creating the quark-antiquark pair and its subsequent hadronization are commonly assumed to be well separated in time and, hence, independent of each other.

Then it proves to be convenient to define the hadronic vacuum polarization tensor (or the two-point correlation function of quark currents, as it is also called)

$$\Pi_{\mu\nu}(q^2) = i \int d^4x \, e^{iqx} \langle 0|T\{J_\mu(x)J_\nu(0)\}|0\rangle = \frac{i}{12\pi^2}(q_\mu q_\nu - g_{\mu\nu}q^2)\Pi(q^2)$$

(1.28)

and isolate its scalar part $\Pi(q^2)$, the latter being the hadronic vacuum polarization function. By virtue of the optical theorem, the hadronic tensor $H_{\mu\nu}$ (1.25) is proportional to the discontinuity of the function $\Pi_{\mu\nu}(q^2)$ across the physical cut that eventually makes it possible to express the total cross section of the process on hand in terms of $\Pi(q^2)$, specifically

$$\sigma\left(e^+e^- \to \text{hadrons}; s\right) = \frac{4}{3}\frac{\alpha_{\text{em}}^2}{s} \, \text{Im} \lim_{\varepsilon \to 0_+} \Pi(s + i\varepsilon).$$

(1.29)

In this equation α_{em} stands for the electromagnetic fine structure constant and s denotes the center-of-mass energy squared. Therefore the R-ratio of electron-positron annihilation into hadrons (1.1) acquires the form

$$R(s) = \frac{1}{\pi} \, \text{Im} \lim_{\varepsilon \to 0_+} \Pi(s + i\varepsilon).$$

(1.30)

As discussed earlier, the experiments on electron-positron annihilation into hadrons play a decisive role in the precise self-consistency tests of QCD and entire Standard Model that puts strong limits on a possible new fundamental physics beyond the latter. This is primarily caused by the fact that the theoretical description of the process on hand, contrary to some other strong interaction processes, does not require the phenomenological models of the process of hadronization. In particular, the summation over hadronic final states in Eq. (1.25) and the completeness of the set of hadronic states (1.26) enable one to eliminate the process of hadronization, which still has no reliable theoretical description, from the analysis. In turn, this opens the way for a model-independent evaluation of the key parameters of the theory.

At the same time, the physical fact that the matrix element of the process on hand (1.22) vanishes identically for the energies below the hadronic production threshold imposes intrinsically nonperturbative constraints on the hadronic tensor $H_{\mu\nu}$ (1.25), as well as on all the related functions. Specifically, this kinematic restriction implies that the hadronic tensor $H_{\mu\nu}$ assumes nonzero

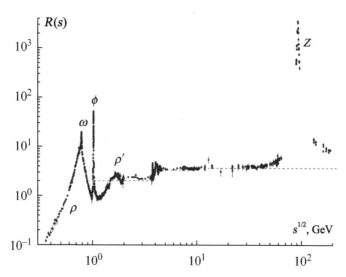

FIG. 1.3 The experimental data on R-ratio of electron-positron annihilation into hadrons (1.1). A *dashed line* corresponds to the parton model prediction $R_{\text{pert}}^{(0)}(s)$ (1.32). The displayed data are available at http://pdg.lbl.gov/2015/hadronic-xsections.

values only for the energies exceeding the hadronic production threshold (i.e., for $q^2 \geq 4m_\pi^2$), whereas $H_{\mu\nu} \equiv 0$ for $q^2 < 4m_\pi^2$, and so does the function $R(s)$ (1.1), namely

$$R(s) \equiv 0, \quad 0 \leq s < 4m_\pi^2. \tag{1.31}$$

Additionally, the aforementioned kinematic restriction implies that the hadronic vacuum polarization function $\Pi(q^2)$ (1.28) has the only cut in the complex q^2-plane, which starts at the hadronic production threshold $q^2 = 4m_\pi^2$ and goes along the positive semiaxis of real q^2, see Fig. 1.4. Basically, this feature also leads to a number of nonperturbative constraints on the related Adler function $D(Q^2)$ (1.46), see Section 1.3 for details. In turn, as it will be discussed in Chapter 4, all these restrictions play a substantial role in the study of the functions on hand at low energies.

The experimental data on R-ratio of electron-positron annihilation into hadrons are displayed in Fig. 1.3. As one can infer from this figure, a distinctive feature of the function $R(s)$ (1.1) is the presence of the resonance peaks, which correspond to the bound states of quarks. The formation of the latter, which involves all the tangled nonperturbative dynamics of colored fields, still has no rigorous unabridged theoretical description. In particular, this fact implies that the direct comparison of the experimental data on R-ratio with its perturbative approximation $R_{\text{pert}}(s)$ is sensible only in the energy intervals, which are located far enough from the hadronic resonances. Alternatively, one can employ the commonly used method of "smearing" the function $R(s)$ [102], which averages out the resonance structure of the R-ratio, see Section 5.2 for details.

TABLE 1.2 Numerical Values of the Factors Appearing in Front of Nonsinglet and Singlet Parts of the Hadronic Vacuum Polarization Function (Second and Third Columns, Respectively)

n_f	$N_c \sum_{f=1}^{n_f} Q_f^2$	$\left(\sum_{f=1}^{n_f} Q_f \right)^2$
1	4/3	4/9
2	5/3	1/9
3	2	0
4	10/3	4/9
5	11/3	1/9
6	5	1

Fig. 1.3 also displays the so-called naive parton model prediction (dashed line)

$$R_{\text{pert}}^{(0)}(s) = N_c \sum_{f=1}^{n_f} Q_f^2, \quad s \to \infty, \tag{1.32}$$

which represents a rather rough approximation of the function $R(s)$ and is valid at high energies only, see Refs. [99, 103]. In particular, in the deep ultraviolet limit the effects due to the masses of all the involved particles, as well as the strong corrections to the process on hand, can be safely neglected. In this case the only distinction between the cross sections of two processes displayed in Figs. 1.1 and 1.2 is accounted for by the product of two factors (1.32). Namely, the factor $N_c = 3$ comes from the fact that every quark can be in three color states, whereas the factors Q_f^2 are due to the difference between the electric charges of the muon and quark of fth flavor. The numerical values of the parton model prediction (1.32), as well as the values of the factor entering the calculation of the singlet part of the hadronic vacuum polarization function at the higher loop levels (see Section 3.2), are listed in Table 1.2. Note that since the factor (1.32) appears in front of all the functions on hand, it will be omitted in what follows. It is also worthwhile to mention that an accurate theoretical analysis of the R-ratio of electron-positron annihilation into hadrons surely requires the proper account of the higher-order (in the strong coupling) contributions to the hadronic vacuum polarization function $\Pi(q^2)$ (1.28). The detailed description of this issue will be given in the next chapters.

It is necessary to emphasize that the results of the QCD perturbation theory are not directly applicable to the study of the strong interaction processes in the timelike domain (see papers [100, 101, 104–108]). In particular, the self-consistent description of such processes (including the electron-positron annihilation into hadrons) can be performed only by making use of the corresponding dispersion relations. For example, the derivation of the parton model prediction (1.32) includes the calculation of the hadronic vacuum polarization function $\Pi^{(0)}(q^2)$ in the zeroth order in the strong coupling (see Eq. 3.32) and the subsequent continuation of the obtained result into the timelike domain (1.40). Equivalently, Eq. (1.32) can also be obtained by making use of the corresponding expression for the Adler function $D^{(0)}(Q^2)$ (see Eqs. 1.46, 3.53) and the inverse relation (1.50). Basically, it appears that the effects due to such continuation of the spacelike perturbative results into the timelike domain substantially affect the function $R(s)$ and thereby play an essential role in the study of the R-ratio of electron-positron annihilation into hadrons and the related quantities. A detailed discussion of this issue will be given in the next two sections as well as in Chapters 4 and 6.

1.2 HADRONIC VACUUM POLARIZATION FUNCTION $\Pi(q^2)$

As noted in the previous section, the kinematics of the process on hand determines the analytic properties of the hadronic vacuum polarization function $\Pi(q^2)$ (1.28) in the complex q^2-plane. Specifically, the physical fact that the production of the final state hadrons in electron-positron annihilation is kinematically forbidden for the energies $q^2 < 4m_\pi^2$ implies that the function $\Pi(q^2)$ has the only cut $q^2 \geq 4m_\pi^2$ along the positive semiaxis of real q^2.

In turn, once the location of the cut of the hadronic vacuum polarization function $\Pi(q^2)$ in the complex q^2-plane is known, one can write down the corresponding dispersion relation by making use of the Cauchy's integral formula (see, e.g., books [109–111] for details):

$$f(x) = \frac{1}{2\pi i} \int_C \frac{f(\xi)}{\xi - x} \, d\xi. \tag{1.33}$$

Since the function $\Pi(q^2)$ logarithmically increases in the ultraviolet asymptotic (see Eq. 3.32), it proves to be convenient to use the once-subtracted Cauchy's integral formula, which directly follows from Eq. (1.33)

$$f(x) = f(x_0) + \frac{1}{2\pi i}(x - x_0) \int_C \frac{f(\xi)}{(\xi - x)(\xi - x_0)} \, d\xi, \tag{1.34}$$

with x_0 being the subtraction point. Thus, for the function $\Pi(q^2)$ (1.28) the Cauchy's integral formula (1.34) subtracted at a point q_0^2 reads

$$\Delta\Pi(q^2, q_0^2) = \frac{1}{2\pi i}(q^2 - q_0^2) \int_C \frac{\Pi(\xi)}{(\xi - q^2)(\xi - q_0^2)} \, d\xi. \tag{1.35}$$

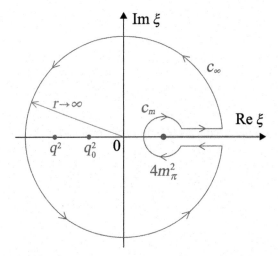

FIG. 1.4 The integration contour C in the complex ξ-plane in the subtracted Cauchy's integral formula (1.35). The physical cut $\xi \geq 4m_\pi^2$ of the hadronic vacuum polarization function $\Pi(\xi)$ (1.28) is shown along the positive semiaxis of real ξ.

In this equation $\Delta\Pi(q^2, q_0^2) = \Pi(q^2) - \Pi(q_0^2)$, whereas the closed integration contour C, which encloses[3] both points q^2 and q_0^2 (commonly, one chooses $q_0^2 = 0$) and goes counterclockwise along the circle of infinitely large radius, is displayed in Fig. 1.4. Because the integration along the circles of infinitely large (c_∞) and infinitesimal (c_m) radii gives no contribution to Eq. (1.35), the hadronic vacuum polarization function can be represented as

$$\Delta\Pi(q^2, q_0^2) = \frac{q^2 - q_0^2}{2\pi i} \lim_{\varepsilon \to 0_+} \left[\int_{4m_\pi^2 + i\varepsilon}^{\infty + i\varepsilon} \frac{\Pi(\xi)\, d\xi}{(\xi - q^2)(\xi - q_0^2)} \right.$$

$$\left. + \int_{\infty - i\varepsilon}^{4m_\pi^2 - i\varepsilon} \frac{\Pi(\xi)\, d\xi}{(\xi - q^2)(\xi - q_0^2)} \right]. \tag{1.36}$$

Then, in Eq. (1.36) it is convenient to change the integration variable $s = \xi - i\varepsilon$ in the first term in the square brackets, which takes the form

$$\int_{4m_\pi^2}^{\infty} \frac{\Pi(s + i\varepsilon)}{(s - q^2)(s - q_0^2)}\, ds + \mathcal{O}(\varepsilon). \tag{1.37}$$

---

3. Since both points q^2 and q_0^2 must be enclosed by the integration contour C in Eq. (1.35), for real q^2 and q_0^2 only the values $q^2 < 4m_\pi^2$ and $q_0^2 < 4m_\pi^2$ can be chosen. The implications of this fact for the massless limit will be discussed in Sections 4.3 and 5.1.

Similarly, after the change of the integration variable $s = \xi + i\varepsilon$ in the second term, the latter reads

$$-\int\limits_{4m_\pi^2}^{\infty} \frac{\Pi(s - i\varepsilon)}{(s - q^2)(s - q_0^2)}\, ds + \mathcal{O}(\varepsilon). \tag{1.38}$$

Hence, the once-subtracted Cauchy's integral formula for $\Pi(q^2)$ (1.35) can be represented as

$$\Delta\Pi(q^2, q_0^2) = \frac{q^2 - q_0^2}{2\pi i}\,\lim_{\varepsilon \to 0_+}\int\limits_{4m_\pi^2}^{\infty} \frac{\Pi(s + i\varepsilon) - \Pi(s - i\varepsilon)}{(s - q^2)(s - q_0^2)}\, ds. \tag{1.39}$$

Recall that for the function $\Pi(\xi)$, which satisfies the condition $\Pi(\xi^*) = \Pi^*(\xi)$, its discontinuity across the physical cut reads

$$R(s) = \frac{1}{\pi}\,\mathrm{Im}\,\lim_{\varepsilon \to 0_+}\Pi(s + i\varepsilon) = \frac{1}{2\pi i}\,\lim_{\varepsilon \to 0_+}[\Pi(s + i\varepsilon) - \Pi(s - i\varepsilon)]. \tag{1.40}$$

Thus, the dispersion relation for the hadronic vacuum polarization function $\Pi(q^2)$ subtracted at a point q_0^2 acquires the well-known form

$$\Delta\Pi(q^2, q_0^2) = (q^2 - q_0^2)\int\limits_{4m_\pi^2}^{\infty} \frac{R(s)}{(s - q^2)(s - q_0^2)}\, ds. \tag{1.41}$$

It is worthwhile to mention that Eqs. (1.40), (1.41) enable one to express the functions $R(s)$ and $\Pi(q^2)$ in terms of each other. Also note that the dispersion relation (1.41) inherently embodies the kinematic restrictions on the process on hand. Namely, Eq. (1.41) explicitly implies that the hadronic vacuum polarization function $\Pi(q^2)$ possesses the only cut[4] along the positive semiaxis of real q^2 starting at the hadronic production threshold $q^2 \geq 4m_\pi^2$, so that its discontinuity (1.40) vanishes everywhere in the complex q^2-plane except for this cut thereby expressing the physical fact that the R-ratio (1.40) assumes nonzero values only for the energies $q^2 \geq 4m_\pi^2$.

Basically, the hadronic vacuum polarization function $\Pi(q^2)$ (1.28), contrary to the R-ratio of electron-positron annihilation into hadrons (1.1), cannot be directly measured in the experiments. Nonetheless, the dispersion relation (1.41) makes it possible to obtain the experimental prediction for the function $\Pi(q^2)$.

---

4. That signifies that the hadronic vacuum polarization function $\Pi(q^2)$ is free of the unphysical singularities.

Specifically, for this purpose the function $R(s)$ entering the integrand of Eq. (1.41) can be approximated by the experimental data on the R-ratio (1.1) at low and intermediate energies and by its perturbative expression[5] at high energies, namely

$$R(s) \simeq \theta(s_0 - s)R_{\text{data}}(s) + \theta(s - s_0)R_{\text{pert}}^{(\ell)}(s). \tag{1.42}$$

In this equation $\theta(x)$ stands for the Heaviside unit step function (1.27), whereas the parameter s_0 of the dimension of mass squared specifies the energy scale of matching[6] of the quantities $R_{\text{data}}(s)$ and $R_{\text{pert}}^{(\ell)}(s)$, and has to be chosen large enough to securely make the perturbative approximation $R_{\text{pert}}^{(\ell)}(s)$ reliable for all $s \geq s_0$. Thus, the experimental prediction for the hadronic vacuum polarization function $\Pi(q^2)$ can be represented in the following form:

$$\Delta\Pi_{\text{exp}}(q^2, q_0^2) = \int_{4m_\pi^2}^{s_0} \frac{(q^2 - q_0^2)R_{\text{data}}(s)}{(s - q^2)(s - q_0^2)} \, ds + \int_{s_0}^{\infty} \frac{(q^2 - q_0^2)R_{\text{pert}}^{(\ell)}(s)}{(s - q^2)(s - q_0^2)} \, ds. \tag{1.43}$$

As one might note, in general, the function $\Delta\Pi_{\text{exp}}(q^2, q_0^2)$ (1.43) depends on the value of parameter s_0, on the loop level ℓ, and (at the higher loop levels) on the employed renormalization scheme.

In fact, the experimental predictions for all the quantities related to the hadronic vacuum polarization function can be obtained in the very same way as for $\Pi(q^2)$ itself. For example, the approximation (1.42) is commonly employed for the evaluation of the Adler function $D(Q^2)$ in the infrared domain (see Section 1.3), as well as for the assessment of the hadronic contributions to such electroweak observables, as the muon anomalous magnetic moment and the shift of the electromagnetic fine structure constant at the scale of Z boson mass, see reviews [112–118] and recent papers [93, 96, 119–121] for the details. It is also worthwhile to mention that for the energies below the mass of τ lepton (i.e., for $s < M_\tau^2$, $M_\tau \simeq 1.777$ GeV [84]) the experimental data on R-ratio of electron-positron annihilation into hadrons can be substituted by the so-called inclusive τ lepton hadronic decay vector spectral function. The original measurements of the latter by ALEPH and OPAL Collaborations, as well as its recent updates can be found in papers [122–125] and [126–131], respectively. However, to make such substitution justified one has to properly account for the effects due to the breaking of the isospin symmetry, see papers [132–134] and references therein for a detailed discussion of this issue.

---

5. The perturbative approximation of R-ratio of electron-positron annihilation into hadrons will be discussed in Section 3.3 and Chapter 6.

6. Not to be confused with the matching condition for the strong running coupling, which will be discussed in Section 2.2.

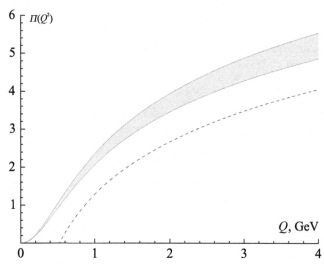

FIG. 1.5 The experimental prediction for the subtracted hadronic vacuum polarization function $\bar{\Pi}_{\exp}(Q^2) = \Delta\Pi_{\exp}(0, -Q^2)$ (Eq. 1.43, *shaded band*) and the pertinent parton model prediction $\Delta\Pi_{\text{pert}}^{(0)}(-Q_0^2, -Q^2)$ (Eq. 1.45, *dashed curve*).

The experimental prediction for the hadronic vacuum polarization function (1.43) (as usual, the subtraction point q_0^2 is set to zero)

$$\bar{\Pi}_{\exp}(Q^2) = \Delta\Pi_{\exp}(0, -Q^2), \quad Q^2 = -q^2 \geq 0 \qquad (1.44)$$

is presented in Fig. 1.5. As one can infer from this figure, the function $\bar{\Pi}_{\exp}(Q^2)$ vanishes in the infrared limit and monotonously increases as the energy scale Q^2 increases. Fig. 1.5 also displays the hadronic vacuum polarization function (1.28) calculated in the leading order of perturbation theory, that is, in the zeroth order in the strong coupling (see Eq. 3.31)

$$\Delta\Pi_{\text{pert}}^{(0)}(-Q_0^2, -Q^2) = \ln\left(\frac{Q^2}{Q_0^2}\right), \qquad (1.45)$$

which corresponds to the naive parton model prediction (1.32). As mentioned in the previous section, the common prefactor $N_c \sum_{f=1}^{n_f} Q_f^2$ is omitted throughout. As one can infer from Fig. 1.5, the function (1.45) tends to its experimental prediction (1.44) at high energies, deviates from the latter at moderate energies, and diverges in the infrared limit $Q^2 \to 0$ irrespective of what value is assigned to the subtraction point Q_0^2. Also note that the perturbative approximation (1.45) cannot be subtracted at the point $Q_0^2 = 0$, which basically originates in the mathematical fact that its analytic properties in the kinematic variable Q^2 differ from those prescribed by the dispersion relation (1.41). The detailed discussion of this issue can be found in Section 5.1. It is also worth mentioning that an accurate theoretical assessment of the hadronic vacuum polarization function $\Pi(q^2)$ (1.28) certainly requires the proper account of the corresponding higher-order perturbative contributions, as well as the relevant nonperturbative constraints, which will be described in the next chapters.

1.3 ADLER FUNCTION $D(Q^2)$

For practical purposes it proves to be particularly convenient to deal with the so-called Adler function[7] $D(Q^2)$, which was proposed by Adler in mid-1970s in paper [100]. This function, being defined as the logarithmic derivative of the hadronic vacuum polarization function (1.28)

$$D(Q^2) = -\frac{d\Pi(-Q^2)}{d\ln Q^2}, \tag{1.46}$$

does not depend on the choice of the subtraction point q_0^2 in Eq. (1.41). It is necessary to outline that the aforementioned kinematic restrictions on the process on hand impose certain intrinsically nonperturbative constraints on the Adler function (1.46), too. As earlier, these constraints are embodied in the corresponding dispersion relation, which directly follows from the definition (1.46) and Eq. (1.41), specifically [100]

$$D(Q^2) = Q^2 \int\limits_{4m_\pi^2}^\infty \frac{R(s)}{(s+Q^2)(s-q_0^2)}\,ds - Q^2 \int\limits_{4m_\pi^2}^\infty \frac{(Q^2+q_0^2)R(s)}{(s+Q^2)^2(s-q_0^2)}\,ds$$

$$= Q^2 \int\limits_{4m_\pi^2}^\infty \frac{R(s)}{(s+Q^2)^2}\,ds. \tag{1.47}$$

In particular, the dispersion relation (1.47) implies that the function $D(Q^2)$ (1.46) possesses the only cut $Q^2 \le -4m_\pi^2$ along the negative semiaxis of real Q^2, that, in turn, signifies that the function $D(Q^2)$ contains no unphysical singularities. Additionally, since the function $R(s)$ (1.1), being a physical observable, acquires finite nonnegative values in the energy range $4m_\pi^2 \le s < \infty$, Eq. (1.47) also implies that the Adler function (1.46) assumes finite nonnegative values in the entire energy range $0 < Q^2 < \infty$ and vanishes in the infrared limit: $D(Q^2) \to 0$ for $Q^2 \to 0$, that, in turn, essentially stabilizes the behavior of the Adler function at low energies. A detailed discussion of the aforementioned constraints and their implications will be given in Chapters 4 and 5.

Basically, the Adler function (1.46) and pertinent dispersion relations play a key role in the theoretical description of the hadron dynamics in the timelike domain. In particular, the QCD perturbation theory and the renormalization group method are directly applicable to the study of the strong interaction

---

7. The Adler function $D(Q^2)$ (1.46) is the function of the spacelike kinematic variable $Q^2 = -q^2 \ge 0$.

processes only in the spacelike (Euclidean) domain. As for the hadron dynamics in the timelike (Minkowskian) domain, for the proper analysis of the latter one first has to relate the results of the QCD perturbation theory with corresponding physical observables. This objective can be achieved in two alternative ways.

Specifically, on the one hand, the dispersion relation (1.47) enables one to obtain the experimental prediction for the Adler function (1.46) in a way identical to that which is described in the previous section, see Eq. (1.52), that constitutes an effective "continuation" of the experimental data on R-ratio into the spacelike domain. On the other hand, one can also continue the results of the QCD perturbation theory into the timelike domain[8] and compare them with relevant hadronic observables. For this purpose, the relation between the functions $D(Q^2)$ and $R(s)$, which is inverse to Eq. (1.47), is required.

The inverse relation, which expresses the function $R(s)$ (1.40) in terms of the Adler function $D(Q^2)$ (1.46) was obtained in early 1980s by Radyushkin [105], Krasnikov and Pivovarov [106]. Specifically, in the definition (1.46) one can separate the variables and integrate the resulting expression in finite limits, that yields

$$-\int_{\Pi(-Q_1^2)}^{\Pi(-Q_2^2)} d\Pi(-Q^2) = \int_{Q_1^2}^{Q_2^2} D(Q^2)\frac{dQ^2}{Q^2}. \tag{1.48}$$

Then, it is convenient to change the integration variable $Q^2 = -\zeta$ on the right-hand side of Eq. (1.48) and set $Q_1^2 = -s - i\varepsilon$ and $Q_2^2 = -s + i\varepsilon$:

$$\Pi(s+i\varepsilon) - \Pi(s-i\varepsilon) = \int_{s+i\varepsilon}^{s-i\varepsilon} D(-\zeta)\frac{d\zeta}{\zeta}. \tag{1.49}$$

Finally, multiplying both sides of this equation by $1/(2\pi i)$, setting $\varepsilon \to 0_+$, and recalling Eq. (1.40) one arrives at the required relation [105, 106]

$$R(s) = \frac{1}{2\pi i} \lim_{\varepsilon \to 0_+} \int_{s+i\varepsilon}^{s-i\varepsilon} D(-\zeta)\frac{d\zeta}{\zeta}. \tag{1.50}$$

The integration contour on the right-hand side of this equation lies in the region of analyticity of the integrand[9] $D(-\zeta)/\zeta$, see Fig. 1.6. The relation (1.50) accounts for the effects due to a continuation of the spacelike perturbative results into the timelike domain and enables one to obtain the explicit form of the function $R(s)$ by making use of the theoretical expression for the Adler

---

8. Discussion of various attempts to continue the results of the QCD perturbation theory into the timelike domain can be found in papers [135–138].

9. Note that the dispersion relation (1.47) implies that $D(-\zeta)/\zeta$ is a regular function at the point $\zeta = 0$.

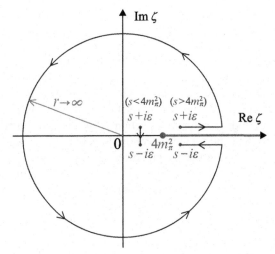

FIG. 1.6 The integration contour in Eq. (1.50). The physical cut $\zeta \geq 4m_\pi^2$ of the Adler function $D(-\zeta)$ (1.46) is shown along the positive semiaxis of real ζ.

function $D(Q^2)$. As it will be discussed in Chapters 4–6, the R-ratio (1.50) drastically differs from the result of naive continuation of the perturbative expression for the Adler function into the timelike domain even in the ultraviolet asymptotic, that, in turn, plays a significant role in the study of the relevant strong interaction processes.

It is also worthwhile to mention that the relation, which expresses the hadronic vacuum polarization function $\Pi(q^2)$ (1.28) in terms of the Adler function $D(Q^2)$ (1.46), was obtained in early 1990s by Pivovarov [139]. In fact, this relation directly follows from Eq. (1.48), specifically

$$\Delta\Pi(-Q^2, -Q_0^2) = -\int_{Q_0^2}^{Q^2} D(\zeta)\frac{d\zeta}{\zeta}, \qquad (1.51)$$

where Q_0^2 and Q^2 stand for the subtraction point and the spacelike kinematic variable, respectively.

Basically, Eqs. (1.40), (1.41), (1.46), (1.47), (1.50), (1.51) constitute the complete set of relations, which express the functions $\Pi(q^2)$, $R(s)$, and $D(Q^2)$ in terms of each other. It is worthwhile to outline that the derivation of the aforementioned relations is based solely[10] on the kinematics of the process on hand and involves neither additional approximations nor phenomenological model-dependent assumptions. At the same time, the derived

---

10. In fact, the derivation of the dispersion relations for the functions on hand also employs a hint on the high-energy behavior of the hadronic vacuum polarization function $\Pi(q^2)$, which prescribes the use of the once-subtracted Cauchy's integral formula (1.34).

relations impose stringent physical intrinsically nonperturbative constraints on the functions $\Pi(q^2)$, $R(s)$, and $D(Q^2)$, which should certainly be accounted for when one comes out of the limits of applicability of perturbation theory. A detailed discussion of this issue will be given in Chapter 4.

Similar to the hadronic vacuum polarization function $\Pi(q^2)$ (1.28), the Adler function $D(Q^2)$ (1.46) cannot be directly measured in the experiments. Nonetheless, the pertinent dispersion relation (1.47) enables one to obtain the experimental prediction for the function $D(Q^2)$ in a way identical to that which is described in the previous section. Specifically, for this purpose the function $R(s)$ appearing in the integrand on the right-hand side of Eq. (1.47) can be approximated by the experimental data on R-ratio of electron-positron annihilation into hadrons (1.1) at low and intermediate energies and by its perturbative expression at high energies, see Eq. (1.42). In this case the dispersion relation for the Adler function (1.47) acquires the following form:

$$D_{\text{exp}}(Q^2) = Q^2 \int\limits_{4m_\pi^2}^{s_0} \frac{R_{\text{data}}(s)}{(s+Q^2)^2} \, ds + Q^2 \int\limits_{s_0}^{\infty} \frac{R_{\text{pert}}^{(\ell)}(s)}{(s+Q^2)^2} \, ds. \tag{1.52}$$

In this equation the parameter s_0 of the dimension of mass squared stands for the energy scale of "matching" of the quantities $R_{\text{data}}(s)$ and $R_{\text{pert}}^{(\ell)}(s)$ (not to be confused with the matching condition for the strong running coupling, which will be delineated in Section 2.2). As one might note the experimental prediction for the Adler function (1.52), in general, depends on the value of the parameter s_0, on the loop level ℓ, and (at the higher loop levels) on the applied renormalization scheme. A detailed discussion of the extraction of the Adler function (1.52) from the experimental data on R-ratio of electron-positron annihilation into hadrons can be found in papers [92, 140, 141].

The experimental prediction for the Adler function $D_{\text{exp}}(Q^2)$ (1.52) is displayed in Fig. 1.7. As one can infer from this figure the function $D_{\text{exp}}(Q^2)$ vanishes in the infrared limit $Q^2 \to 0$, increases as the energy scale Q^2 increases, and approaches a constant value in the ultraviolet asymptotic $Q^2 \to \infty$. The latter is determined by the perturbative approximation of the Adler function calculated in the zeroth order in the strong coupling (see Eq. 3.53)

$$D_{\text{pert}}^{(0)}(Q^2) = 1, \quad Q^2 \to \infty, \tag{1.53}$$

which is shown by dashed line in Fig. 1.7. The expression (1.53) corresponds to the parton model predictions for the hadronic vacuum polarization function (1.45) and the R-ratio of electron-positron annihilation into hadrons (1.32). As noted in the previous sections, the common prefactor $N_c \sum_{f=1}^{n_f} Q_f^2$ is omitted throughout. It is also worthwhile to mention that an accurate theoretical

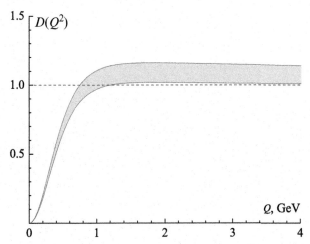

FIG. 1.7 The experimental prediction for the Adler function $D_{\text{exp}}(Q^2)$ (Eq. 1.52, *shaded band*) and its parton model prediction $D_{\text{pert}}^{(0)}(Q^2)$ (Eq. 1.53, *dashed line*).

description of the Adler function $D(Q^2)$ (1.46) evidently requires the proper account of the corresponding higher-order perturbative contributions, as well as the pertinent nonperturbative constraints, which will be discussed in the next chapters.

Chapter 2

Perturbative QCD Invariant Charge

The invariant charge of Quantum Chromodynamics (QCD) $\alpha_s(Q^2)$, being the measure of strength of the strong interaction between colored objects, plays a fundamental role in elementary particle physics. In particular, the theoretical study of a variety of the hadronic observables and the profound analysis of the corresponding experimental data, being inherently based on the strong running coupling $\alpha_s(Q^2)$, constitutes a decisive self-consistency test of QCD and the entire Standard Model, and puts strict limits on a possible new physics beyond the latter.

2.1 RENORMALIZATION GROUP EQUATION

The theoretical study of a process in particle physics being performed beyond the leading order of perturbation theory entails the divergence of the corresponding loop diagrams. The latter cannot be directly related to experimentally measurable characteristics of the process on hand. Nonetheless, in the renormalizable theories the bare quantities entering Lagrangian can be redefined, or "renormalized," in the way that makes the theoretical expression for a physical observable finite at any given order of perturbation theory. At the same time the renormalization procedure inevitably brings into the theory an additional parameter of the dimension of mass, namely, the renormalization scale μ. The dependence of the renormalized quantities on this scale is governed by the renormalization group equation. The detailed description of the issue on hand can be found in the basic textbooks on Quantum Field Theory as well as in a number of extensive reviews [142–152].

For the strong running coupling $\alpha_s(\mu^2) = g^2(\mu^2)/(4\pi)$ the corresponding renormalization group equation in the differential form reads

$$\frac{d\ln\left[g^2(\mu^2)\right]}{d\ln\mu^2} = \beta\left[g(\mu^2)\right]. \tag{2.1}$$

Strong Interactions in Spacelike and Timelike Domains. http://dx.doi.org/10.1016/B978-0-12-803439-2.00002-8

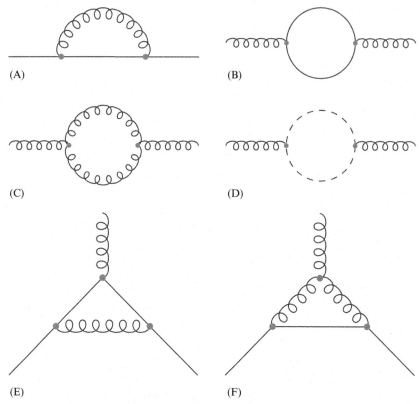

(A)

(B)

(C)

(D)

(E)

(F)

FIG. 2.1 The one-loop QCD corrections to the quark propagator (*plot A*), to the gluon propagator (*plots B–D*), and to the quark-gluon vertex (*plots E and F*).

In the framework of perturbation theory the β function appearing on the right-hand side of this equation can be approximated by the power series in $g^2(\mu^2)$

$$\beta\left[g(\mu^2)\right] \simeq -\left\{\beta_0\left[\frac{g^2(\mu^2)}{16\pi^2}\right] + \beta_1\left[\frac{g^2(\mu^2)}{16\pi^2}\right]^2 + \beta_2\left[\frac{g^2(\mu^2)}{16\pi^2}\right]^3 + \cdots\right\},$$

(2.2)

the smallness of $g^2(\mu^2)$ being assumed. At present the explicit form of only the first five coefficients of the perturbative expansion of QCD β function are known. In particular, the one-loop coefficient β_0, which involves the calculation of the diagrams displayed in Fig. 2.1, was obtained in 1973 in the milestone works by Gross and Wilczek [16] and Politzer [17]

$$\beta_0 = 11 - \frac{2}{3}n_f,$$

(2.3)

see also papers [15, 20, 21] by 't Hooft. This result, which implies that the first term of the β function (2.2) is negative,[1] led to the discovery of the asymptotic freedom in QCD and paved the way for the description of the strong interaction processes at high energies within the perturbative approach.

The calculation of coefficients β_j at the higher loop levels becomes more elaborate due to the increasing number of involved diagrams. For example, the calculation of the two-loop coefficient

$$\beta_1 = 102 - \frac{38}{3} n_f \tag{2.4}$$

was performed by Caswell [153] and Jones [154] just about a year after the appearance of the one-loop result and was repeated a few years later by Egorian and Tarasov [155]. The expression for the three-loop coefficient β_2 was obtained for the first time in 1980 by Tarasov, Vladimirov, and Zharkov [156] and was rederived by Larin and Vermaseren about a decade later [157]:

$$\beta_2 = \frac{2857}{2} - \frac{5033}{18} n_f + \frac{325}{54} n_f^2. \tag{2.5}$$

The calculation of the four-loop coefficient β_3, which involves about 50000 diagrams, was performed only in late 1990s in papers by van Ritbergen, Vermaseren, and Larin [158] and Chetyrkin, Kniehl, and Steinhauser [159]:

$$\beta_3 = \frac{149753}{6} + 3564\zeta(3) - n_f \left[\frac{1078361}{162} + \frac{6508}{27}\zeta(3) \right]$$
$$+ n_f^2 \left[\frac{50065}{162} + \frac{6472}{81}\zeta(3) \right] + n_f^3 \frac{1093}{729}. \tag{2.6}$$

The expression for the five-loop coefficient β_4 was obtained only two decades later by Baikov, Chetyrkin, and Kuhn [160]:

$$\beta_4 = \frac{8157455}{16} + \frac{621885}{2}\zeta(3) - \frac{88209}{2}\zeta(4) - 288090\zeta(5)$$
$$+ n_f \left[-\frac{336460813}{1944} - \frac{4811164}{81}\zeta(3) + \frac{33935}{6}\zeta(4) + \frac{1358995}{27}\zeta(5) \right]$$
$$+ n_f^2 \left[\frac{25960913}{1944} + \frac{698531}{81}\zeta(3) - \frac{10526}{9}\zeta(4) - \frac{381760}{81}\zeta(5) \right]$$
$$+ n_f^3 \left[-\frac{630559}{5832} - \frac{48722}{243}\zeta(3) + \frac{1618}{27}\zeta(4) + \frac{460}{9}\zeta(5) \right]$$
$$+ n_f^4 \left[\frac{1205}{2916} - \frac{152}{81}\zeta(3) \right]. \tag{2.7}$$

---

1. Formally, this statement is valid for $n_f \leq 16$. At present, only six quark flavors have been discovered, whereas the experimental searches have produced no clear evidence for the existence of the fourth-generation quarks.

The explicit form of the higher-order coefficients of the QCD β function perturbative expansion (2.2) is currently unavailable.

In Eqs. (2.3)–(2.7) n_f denotes the number of active flavors (see later), whereas $\zeta(x)$ stands for the Riemann ζ function,

$$\zeta(x) = \sum_{n=1}^{\infty} \frac{1}{n^x}, \tag{2.8}$$

$\zeta(3) \simeq 1.20206$, $\zeta(4) = \pi^4/90 \simeq 1.08232$, and $\zeta(5) \simeq 1.03693$. The coefficients (2.3)–(2.7) correspond to the SU(3) gauge group. The first two coefficients of the QCD β function perturbative expansion (2.3), (2.4) are universal (in mass-independent renormalization schemes), whereas the higher-loop coefficients are scheme-dependent. In particular, the expressions (2.5)–(2.7) correspond to the "modified minimal subtraction scheme" [161], also known as the "$\overline{\text{MS}}$-scheme," which will be briefly recounted in Section 3.1. In what follows the scheme-dependent coefficients β_j are assumed to be taken in the $\overline{\text{MS}}$-scheme unless otherwise specified.

There is a number of hints, which might shed some light on the uncalculated coefficients β_j. For example, the so-called "large-n_f expansion" method (the technique was initially developed in papers [162–167]) makes it possible to derive the term of a perturbative coefficient, which is proportional to the highest power of n_f. Specifically, in mid-1990s such terms of the coefficients β_j have been obtained by Gracey [168] at the four-loop and five-loop levels. These were unavailable at that time and appear to be identical to the last terms of Eqs. (2.6), (2.7), respectively, calculated afterward. Additionally, Gracey obtained [168] the corresponding expressions at the six-loop level

$$\beta_5 = n_f^5 \left[\frac{2069}{10935} + \frac{1040}{729}\zeta(3) - \frac{608}{405}\zeta(4) \right] + \mathcal{O}\left(n_f^4\right) \tag{2.9}$$

and seven-loop level

$$\beta_6 = n_f^6 \left[\frac{349}{4374} + \frac{8744}{19683}\zeta(3) + \frac{2600}{2187}\zeta(4) - \frac{1216}{729}\zeta(5) \right] + \mathcal{O}\left(n_f^5\right). \tag{2.10}$$

It is also worthwhile to mention that the numerical values of the higher-order perturbative coefficients may be estimated by making use of the so-called Padé approximants (see, in particular, papers [169–171] and references therein).

It is necessary to outline that in the calculations of the coefficients of the QCD β function perturbative expansion (2.3)–(2.7) the quark masses m_f have been neglected. Basically, this assumption is sensible only if the value of the characteristic energy scale of a process on hand is bound to certain kinematic intervals. Specifically, if, on the one hand, the energy of the studied process substantially exceeds the mass of quark of fth flavor, then all the quarks of first f flavors contribute to the pertinent observables and their masses can be safely

TABLE 2.1 Numerical Values of the β Function Perturbative Expansion Coefficients β_j up to the Five-Loop Level ($0 \leq j \leq 4$) (See Eqs. 2.3–2.7)

n_f	β_0	$\beta_1 \times 10^{-1}$	$\beta_2 \times 10^{-2}$	$\beta_3 \times 10^{-3}$	$\beta_4 \times 10^{-4}$
0	11.0000	10.2000	14.2850	29.2430	53.7148
1	10.3333	8.9333	11.5491	22.7033	36.8320
2	9.6667	7.6667	8.9335	16.9827	23.3215
3	9.0000	6.4000	6.4383	12.0904	13.0378
4	8.3333	5.1333	4.0635	8.0352	5.8311
5	7.6667	3.8667	1.8091	4.8262	1.5471
6	7.0000	2.6000	−0.3250	2.4723	0.0271

Notes: The three-loop, four-loop, and five-loop coefficients (β_2, β_3, and β_4) correspond to the $\overline{MS}$-scheme.

set to zero. On the other hand, if the energy of the studied process is substantially less than the mass of quark of $(f + 1)$th flavor, then, in accordance with the theorem elaborated by Symanzik [172], Appelquist and Carazzone [173], the quarks of $(f + 1)$th and all subsequent flavors do not contribute to the relevant physical observables. In other words, heavy quarks "decouple" from the low-energy physics and their existence, to a certain extent, can be merely ignored (see the discussion of this issue in papers [174–177] and references therein for details). Thus one arrives at a rather simplified approximation of QCD by an effective theory, which retains only the quarks of first f flavors and assumes them all to be massless. In this case, f specifies the so-called number of active flavors n_f, which is the piecewise continuous function of the kinematic variable

$$n_f(Q^2) = \sum_{j=1}^{6} \theta\left(Q^2 - m_j^2\right), \qquad (2.11)$$

with $\theta(x)$ being the Heaviside unit step function (1.27). It is worthwhile to note here that when the energy scale of a process on hand overpasses a quark mass threshold, the number of active flavors n_f (2.11) changes its value abruptly, and so do all the involved perturbative coefficients, since they explicitly depend on n_f. In turn, this fact implies that certain quantities, such as the QCD scale parameter Λ, which governs the running of the strong coupling, have also to be changed accordingly at this threshold, which, in turn, constitutes the so-called matching procedure. The description of the latter will be given in the next section.

For practical purposes it proves to be particularly convenient to represent the renormalization group equation for the perturbative strong running coupling (2.1) in terms of the so-called QCD "couplant" $a_s(\mu^2) \equiv$

TABLE 2.2 Numerical Values of the Coefficients $B_j = \beta_j/\beta_0^{j+1}$ (2.12) up to the Five-Loop Level ($0 \leq j \leq 4$)

n_f	B_0	B_1	B_2	B_3	B_4
0	1.0000	0.8430	1.0733	1.9973	3.3353
1	1.0000	0.8366	1.0467	1.9913	3.1262
2	1.0000	0.8205	0.9890	1.9449	2.7629
3	1.0000	0.7901	0.8832	1.8428	2.2080
4	1.0000	0.7392	0.7022	1.6662	1.4510
5	1.0000	0.6578	0.4015	1.3969	0.5841
6	1.0000	0.5306	−0.0948	1.0297	0.0161

Notes: The three-loop, four-loop, and five-loop coefficients (B_2, B_3, and B_4) correspond to the $\overline{MS}$-scheme.

$\alpha_s(\mu^2)\beta_0/(4\pi) = g^2(\mu^2)\beta_0/(16\pi^2)$. Specifically, with this notation Eq. (2.1) takes the following form at the ℓ-loop level

$$\frac{d\ln\left[a_s^{(\ell)}(\mu^2)\right]}{d\ln\mu^2} = -\sum_{j=0}^{\ell-1} B_j \left[a_s^{(\ell)}(\mu^2)\right]^{j+1}, \quad B_j = \frac{\beta_j}{\beta_0^{j+1}}, \tag{2.12}$$

the perturbative approximation for the β function (2.2) being employed. In what follows the superscript in brackets "(ℓ)" specifies the loop level and not to be confused with ℓth derivative of the function on hand. It is worth mentioning here that for the physically admissible values of the number of active flavors (i.e., for $0 \leq n_f \leq 6$) the available coefficients of the β function perturbative expansion (2.3)–(2.7) assume rather large values, whereas the coefficients B_j (2.12) do not. Also note that for $0 \leq n_f \leq 6$ and $0 \leq j \leq 4$ all the coefficients β_j (2.3)–(2.7) and B_j (2.12) acquire positive values except for the three-loop coefficients β_2 and B_2, which assume negative values for $n_f = 6$ (see Tables 2.1 and 2.2 for details). The solution to Eq. (2.12) at various loop levels will be addressed in Sections 2.2 and 2.3 as well as in Appendix A.

2.2 ONE-LOOP PERTURBATIVE QCD RUNNING COUPLING

At the one-loop level the renormalization group equation (2.12) for the QCD invariant charge $\alpha_s(Q^2)$ assumes a quite simple form, namely

$$\frac{d\ln\left[a_s^{(1)}(\mu^2)\right]}{d\ln\mu^2} = -B_0 a_s^{(1)}(\mu^2), \tag{2.13}$$

where $a_s^{(1)}(\mu^2) = \alpha_s^{(1)}(\mu^2)\beta_0/(4\pi)$ stands for the corresponding couplant and $B_0 = 1$ by definition. This equation can easily be integrated explicitly. Specifically, it is convenient to separate the variables

$$-\frac{da_s^{(1)}(\mu^2)}{\left[a_s^{(1)}(\mu^2)\right]^2} = d\ln\mu^2 \tag{2.14}$$

and then integrate Eq. (2.14) in finite limits, that eventually leads to the following relation between the values of the one-loop QCD couplant at a reference energy scale Q_0^2 and at an arbitrary[2] energy scale Q^2:

$$\frac{1}{a_s^{(1)}(Q^2)} - \frac{1}{a_s^{(1)}(Q_0^2)} = \ln\left(\frac{Q^2}{Q_0^2}\right). \tag{2.15}$$

This equation determines the one-loop perturbative strong running coupling $\alpha_s^{(1)}(Q^2)$ and allows one to express it as a function of Q^2, Q_0^2, and $\alpha_s^{(1)}(Q_0^2)$, namely

$$\alpha_s^{(1)}(Q^2) = \frac{\alpha_s^{(1)}(Q_0^2)}{1 + \alpha_s^{(1)}(Q_0^2)\ln(Q^2/Q_0^2)\beta_0/(4\pi)}. \tag{2.16}$$

It is necessary to emphasize once again that the perturbation theory and the renormalization group method are applicable only in the spacelike domain $Q^2 = -q^2 > 0$.

Basically, the QCD invariant charge in the form of Eq. (2.16) can be employed in all practical applications. At the same time one commonly prefers to deal with a single input parameter instead of the pair of reference values $\left[Q_0^2, \alpha_s(Q_0^2)\right]$, which constitute the boundary condition for Eq. (2.12). For this purpose in Eq. (2.15) all the terms independent of the energy scale Q^2 and of the value of the QCD invariant charge at Q^2 can be absorbed into the so-called QCD scale parameter Λ of the dimension of mass,[3] specifically

$$\frac{1}{a_s^{(1)}(Q^2)} = \ln\left(\frac{Q^2}{Q_0^2}\right) + \frac{1}{a_s^{(1)}(Q_0^2)} \equiv \ln\left(\frac{Q^2}{\Lambda^2}\right), \tag{2.17}$$

where

$$\Lambda^2 = Q_0^2 \exp\left[-\frac{4\pi}{\beta_0}\frac{1}{\alpha_s^{(1)}(Q_0^2)}\right]. \tag{2.18}$$

---

2. It is assumed that both energy scales Q_0^2 and Q^2 belong to a kinematic interval with the same number of active flavors n_f (see discussion of this issue below).

3. It is worthwhile to note that the QCD scale parameter Λ appears even in the massless limit of the theory.

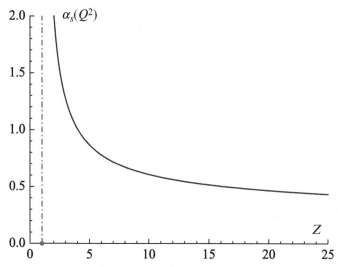

FIG. 2.2 The one-loop perturbative QCD invariant charge (Eq. 2.19, *solid curve*). The values of parameters: $n_f = 3$ active flavors, $z = Q^2/\Lambda^2$. The location of the unphysical "ghost pole" ($Q^2 = \Lambda^2$) is marked by the *vertical dot-dashed line*.

In this case the solution of the renormalization group equation for the one-loop QCD invariant charge (2.12) acquires the well-known compact form:

$$\alpha_s^{(1)}(Q^2) = \frac{4\pi}{\beta_0} \frac{1}{\ln z}, \quad z = \frac{Q^2}{\Lambda^2}. \tag{2.19}$$

Obviously, the expression for the one-loop strong running coupling (2.16) is equivalent to the pair of expressions (2.18), (2.19) and vice versa.

Fig. 2.2 displays the one-loop perturbative QCD invariant charge (2.19). As one can infer from this figure, the strong running coupling $\alpha_s^{(1)}(Q^2)$ is the decreasing function of its argument that is caused by the negative value of the first term of perturbative expansion of the renormalization group β function (2.2) (i.e., $\beta_0 > 0$). As mentioned earlier this fact implements the asymptotic freedom of QCD, which consists in diminishing the strength of the strong interaction between colored objects with the increasing characteristic energy of a physical process on hand. In turn this distinctive feature of QCD enables one to describe various strong interaction processes, if the corresponding energy scale is high enough, within the perturbative approach. In this case a physical observable can be approximated by the perturbative power series in the strong running coupling $\alpha_s(Q^2)$, the latter being the small expansion parameter.

However, it is necessary to emphasize that the QCD perturbation theory is of a limited applicability range since the invariant charge diverges at low energies. In particular, at $Q^2 \simeq \Lambda^2$ the one-loop expression (2.19) can be represented as

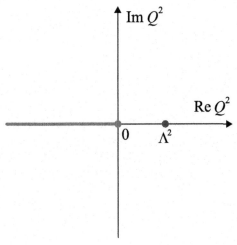

FIG. 2.3 The singularities of the one-loop perturbative strong running coupling $\alpha_s^{(1)}(Q^2)$ (2.19) in the complex Q^2-plane: cut $Q^2 \le 0$ along the negative semiaxis of real Q^2 and unphysical singularity (pole of first order) at $Q^2 = \Lambda^2$.

$$\alpha_s^{(1)}(Q^2) \simeq \frac{4\pi}{\beta_0} \left(\frac{\Lambda^2}{Q^2 - \Lambda^2} + \frac{1}{2} - \frac{1}{12} \frac{Q^2 - \Lambda^2}{\Lambda^2} \right), \quad Q^2 \to \Lambda^2. \quad (2.20)$$

The singularity structure of the perturbative strong running coupling $\alpha_s^{(1)}(Q^2)$ in the complex Q^2-plane is shown in Fig. 2.3. Specifically, the one-loop QCD invariant charge (2.19) contains the cut in the kinematic variable Q^2 along the negative semiaxis of real Q^2 (which can be called physical if the masses of all the involved particles are set to zero) and the unphysical singularity at $Q^2 = \Lambda^2$ (namely, the pole of first order), which is called "ghost pole" or "Landau pole." Evidently, the presence of unphysical singularities contradicts the general principles of the theory, is inconsistent with the dispersion relations discussed in Chapter 1, and makes the QCD perturbation theory inapplicable to the study of hadron dynamics at low energies.

The appearance of unphysical singularities in the solution to the renormalization group equation for the strong running coupling (2.1) is primarily caused by the use of perturbative approximation of the corresponding β function (2.2). In particular, the latter is only valid for small values of the QCD invariant charge $\alpha_s(Q^2)$, that takes place only in the high energy region $Q^2 \gg \Lambda^2$. At low energies the theory comes out of the small coupling regime, the approximation (2.2) becomes inapplicable, and the resulting solution to the corresponding renormalization group equation loses the physical sense. In practice the QCD perturbation theory is capable of describing the strong interaction processes for $Q \gtrsim 1.5$ GeV. To perform the theoretical analysis of hadron dynamics at lower energies one inevitably resorts to a variety of the nonperturbative methods.

It is worthwhile to mention here that the unphysical singularities also appear in the perturbative QED calculations. Specifically, the perturbative QED invariant charge contains unphysical singularities in the asymptotic ultraviolet region, which also contradicts the basic principles of the theory. The construction of the nonperturbative QED running coupling, which is free of unphysical singularities, was discussed in, for example, papers [178–180]. Nevertheless, in Quantum Electrodynamics (contrary to QCD) the perturbation theory fails at the energy scales of the order of the Planck mass $M_{Pl} \simeq 1.22 \times 10^{19}$ GeV. Therefore, since the highest energy achieved in contemporary particle physics experiments is incommensurably less than M_{Pl}, the theoretical analysis of QED observables in fact does not require any nonperturbative method and can reliably be performed in the framework of perturbative approach.

It is necessary to emphasize that the strong running coupling $\alpha_s(Q^2)$ cannot be directly measured in the experiments. Nonetheless, as mentioned earlier, certain measurable characteristics of various hadronic processes, such as, for example, pertinent cross sections and decay widths, can be parameterized in terms of $\alpha_s(Q^2)$ in the framework of perturbation theory. In turn, this allows one to evaluate the QCD invariant charge $\alpha_s(Q^2)$ at a given energy scale by making use of the corresponding experimental data. For example, let us assume that a physical observable $H(Q^2)$ can be approximated within perturbation theory by

$$H_{\text{pert}}^{(\ell)}(Q^2) = 1 + \sum_{j=1}^{\ell} h_j \left[\alpha_s^{(\ell)}(Q^2) \right]^j, \quad Q^2 \to \infty, \tag{2.21}$$

where ℓ specifies the loop level and h_j stand for the relevant perturbative expansion coefficients. Then, let us also assume that at the energy scale Q_0^2, which supports the validity of perturbative approximation (2.21), the value of the QCD observable H was measured experimentally:

$$H(Q_0^2) = H_0, \tag{2.22}$$

the uncertainty of measurement being neglected for the sake of simplicity. These two inputs, namely, the theoretical expression (2.21) and the experimental value (2.22), enable one to evaluate the ℓ-loop QCD invariant charge at the provided reference scale, $\alpha_s^{(\ell)}(Q_0^2)$, or, alternatively, to obtain the value of corresponding QCD scale parameter Λ. For instance, at the one-loop level (i.e., for $\ell = 1$) this prescription yields, respectively,

$$\alpha_s^{(1)}(Q_0^2) = \frac{1}{h_1}(H_0 - 1) \tag{2.23}$$

and

$$\Lambda^2 = Q_0^2 \exp\left(-\frac{4\pi}{\beta_0}\frac{h_1}{H_0 - 1}\right). \tag{2.24}$$

In turn, the value of the one-loop strong running coupling $\alpha_s^{(1)}(Q^2)$ can be found at an arbitrary[4] energy scale Q^2 in two equivalent ways. Specifically, for this purpose one can either use Eq. (2.16) with the provided value of Q_0^2 and evaluated $\alpha_s^{(1)}(Q_0^2)$ (2.23) or employ Eq. (2.19) with the obtained value of the QCD scale parameter (2.24). The latter option, being essentially the same as the former one but appearing to be technically less demanding, is commonly used in the majority of practical applications.

At the higher loop levels (i.e., for $\ell \geq 2$) the prescription described earlier can be implemented in the very same way as at the one-loop level. At the same time one has to be aware that, in general, at different loop levels this prescription returns different values for the QCD invariant charge at the reference energy scale, $\alpha_s^{(\ell)}(Q_0^2)$, as well as different values for the QCD scale parameter Λ. Besides, at the higher loop levels the obtained values of $\alpha_s^{(\ell)}(Q_0^2)$ and Λ depend on the applied subtraction scheme, because the involved perturbative coefficients h_j and β_j become scheme-dependent. It is also worthwhile to mention that the foregoing remarks imply that a confident experimental verification of the self-consistency of the perturbative description of the strong interaction processes requires at least two measurements of a pertinent observable at the separate energy scales or at least two measurements of different QCD observables at arbitrary energy scales. The evaluation of the QCD invariant charge $\alpha_s(Q^2)$ at the higher loop levels proceeding from the experimental measurements of various observables, as well as the assessment of the world average value of the strong running coupling at the scale of Z boson mass $\alpha_s(M_Z^2)$ and the corresponding value of the scale parameter Λ, are thoroughly discussed in a number of reviews devoted to the study of the strong interaction processes in the framework of perturbative approach (see, in particular, Refs. [84, 181–184]).

It is important to stress that the prescription delineated earlier allows one to find the value of the strong running coupling $\alpha_s(Q^2)$ only within the energy range corresponding to the number of active flavors n_f determined by the provided reference scale Q_0^2. In other words the values of QCD invariant charge $\alpha_s(Q_0^2)$ and the scale parameter Λ obtained for a given n_f cannot be directly used in the kinematic intervals, which correspond to other values of n_f. In particular, this is caused by the fact that the involved perturbative expansion coefficients, being dependent on the number of active flavors n_f, change their values abruptly at the quark mass thresholds $Q^2 = m_f^2$ (see Eq. 2.11). Therefore the pair of values $\left[Q_0^2, \alpha_s(Q_0^2)\right]$, as well as the value of the QCD scale parameter Λ, also must be altered accordingly at these thresholds. To relate the effective theories, which correspond to n_f and to $(n_f + 1)$ active flavors, one commonly requires their mutual consistency at the $(f + 1)$th flavor threshold $Q^2 = m_{f+1}^2$.

---

4. Within the kinematic interval corresponding to the number of active flavors n_f specified by Q_0^2 (see the discussion below).

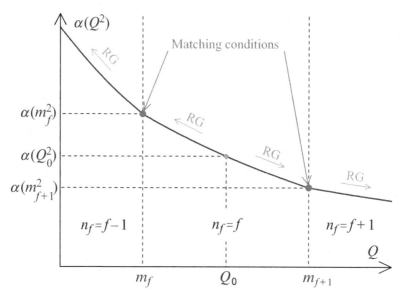

FIG. 2.4 Schematic representation of the matching procedure for the strong running coupling $\alpha_s(Q^2)$ at quark mass thresholds.

Eventually, this results in a rather nontrivial matching condition for the running couplings of those effective theories. Specifically, at the one-loop and two-loop levels the matching condition for the QCD invariant charge $\alpha_s(Q^2)$ assumes a native form of its continuity at the $(f+1)$th flavor threshold (see Eq. 2.25 and Fig. 2.4), whereas at the higher loop levels the strong running coupling becomes discontinuous at the flavor thresholds (see papers [174–177] and references therein).

To illustrate this issue in more detail let us look at the matching procedure at the one-loop level. As earlier, let us assume that at a reference energy scale Q_0^2, which corresponds to $n_f = f$ active flavors (i.e., $m_f^2 \leq Q_0^2 \leq m_{f+1}^2$), the experimental measurement of a physical observable has been performed and the respective values[5] of the strong running coupling $\alpha_{[f]}^{(1)}(Q_0^2)$ and the scale parameter $\Lambda_{[f]}$ have been found in the way described earlier (see Eqs. 2.23, 2.24). Then to properly continue the QCD invariant charge into the kinematic interval corresponding to, for example, $n_f = f + 1$ active flavors one first has to obtain the demarcating value of the strong running coupling at the respective threshold $\alpha_{[f]}^{(1)}(m_{f+1}^2)$, which can easily be achieved by making use of either Eq. (2.16) or (2.19).

---

5. In the rest of this section the index in square brackets specifies the used number of active flavors. For example, the notation $\Lambda_{[f]}$ indicates that this scale parameter is evaluated for $n_f = f$.

To step into the neighboring kinematic interval one has to employ the matching condition for the strong running coupling $\alpha_s^{(1)}(Q^2)$. At the one-loop level this condition takes a quite simple form, which merely corresponds to the continuity of $\alpha_s^{(1)}(Q^2)$ across the quark mass threshold:

$$\alpha_{[f]}^{(1)}\left(m_{f+1}^2\right) = \alpha_{[f+1]}^{(1)}\left(m_{f+1}^2\right). \tag{2.25}$$

It is worthwhile to mention here that the condition of the form of Eq. (2.25), in general, does not provide continuity for the perturbative approximation of a physical observable on hand (2.21) at the threshold $Q^2 = m_{f+1}^2$. Additionally, the derivatives of the resulting strong running coupling also appear to be discontinuous at the threshold.

Then in the kinematic interval corresponding to $n_f = f + 1$ active flavors, the pair of values $\left[m_{f+1}^2, \alpha_{[f+1]}^{(1)}\left(m_{f+1}^2\right)\right]$ constitutes the boundary condition for the one-loop differential equation (2.14). The latter enables one to evaluate the QCD invariant charge at an arbitrary energy scale within this interval (namely, for $m_{f+1}^2 \leq Q^2 \leq m_{f+2}^2$). In particular, this can be performed by making use of Eq. (2.16), which acquires the following form:

$$\alpha_{[f+1]}^{(1)}(Q^2) = \frac{\alpha_{[f]}^{(1)}\left(m_{f+1}^2\right)}{1 + \alpha_{[f]}^{(1)}\left(m_{f+1}^2\right)\ln\left(Q^2/m_{f+1}^2\right)\beta_0^{[f+1]}/(4\pi)}. \tag{2.26}$$

As mentioned earlier one can equivalently find the corresponding value of the QCD scale parameter $\Lambda_{[f+1]}$ (2.18) and then evaluate the strong running coupling by making use of Eq. (2.19), specifically

$$\alpha_{[f+1]}^{(1)}(Q^2) = \frac{4\pi}{\beta_0^{[f+1]}}\frac{1}{\ln\left(Q^2/\Lambda_{[f+1]}^2\right)},$$

$$\Lambda_{[f+1]}^2 = m_{f+1}^2 \exp\left[-\frac{4\pi}{\beta_0^{[f+1]}}\frac{1}{\alpha_{[f]}^{(1)}\left(m_{f+1}^2\right)}\right]. \tag{2.27}$$

It is worth noting that the explicit expression for the QCD invariant charge (2.19) along with the matching condition (2.25) also enable one to directly relate the one-loop scale parameters Λ corresponding to $n_f = f$ and $n_f = f + 1$ active flavors to each other, namely

$$\Lambda_{[f+1]} = m_{f+1}\left(\frac{\Lambda_{[f]}}{m_{f+1}}\right)^{\beta_0^{[f]}/\beta_0^{[f+1]}}. \tag{2.28}$$

Note that the continuation of the strong running coupling $\alpha_s(Q^2)$ from the kinematic interval corresponding to $n_f = f$ active flavors into the neighboring kinematic interval with $n_f = f - 1$ active flavors can be performed in a very similar way. Specifically, for $m_{f-1}^2 \leq Q^2 \leq m_f^2$ one can either use the QCD invariant charge in the form of Eq. (2.16), which reads

$$\alpha_{[f-1]}^{(1)}(Q^2) = \frac{\alpha_{[f]}^{(1)}\left(m_f^2\right)}{1 + \alpha_{[f]}^{(1)}\left(m_f^2\right) \ln\left(Q^2/m_f^2\right)\beta_0^{[f-1]}/(4\pi)}, \tag{2.29}$$

or evaluate the scale parameter $\Lambda_{[f-1]}$ (2.18) and employ expression (2.19), namely

$$\alpha_{[f-1]}^{(1)}(Q^2) = \frac{4\pi}{\beta_0^{[f-1]}} \frac{1}{\ln\left(Q^2/\Lambda_{[f-1]}^2\right)}, \quad \Lambda_{[f-1]}^2 = m_f^2 \exp\left[-\frac{4\pi}{\beta_0^{[f-1]}} \frac{1}{\alpha_{[f]}^{(1)}\left(m_f^2\right)}\right]. \tag{2.30}$$

At the same time, one can also directly relate the QCD scale parameters corresponding to $n_f = f$ and $n_f = f - 1$ active flavors likewise Eq. (2.28):

$$\Lambda_{[f-1]} = m_f \left(\frac{\Lambda_{[f]}}{m_f}\right)^{\beta_0^{[f]}/\beta_0^{[f-1]}}. \tag{2.31}$$

At the higher loop levels ($\ell \geq 2$) the matching procedure for the QCD invariant charge $\alpha_s^{(\ell)}(Q^2)$ remains conceptually the same as at the one-loop level (see Fig. 2.4). Specifically, one first finds the value of the ℓ-loop strong running coupling at a reference energy scale $\alpha_{[f]}^{(\ell)}(Q_0^2)$, or evaluates the corresponding scale parameter $\Lambda_{[f]}$, by making use of an additional input, such as the pertinent experimental data. Then by employing the solution to the ℓ-loop perturbative renormalization group equation (2.12), which will be discussed in the next section, one obtains the value of the QCD invariant charge $\alpha_{[f]}^{(\ell)}\left(m_{f+1}^2\right)$ at the boundary of the given kinematic interval $m_f^2 \leq Q_0^2 \leq m_{f+1}^2$. After that to find the value of the strong running coupling at the corresponding boundary of the neighboring kinematic interval $m_{f+1}^2 \leq Q^2 \leq m_{f+2}^2$, namely, $\alpha_{[f+1]}^{(\ell)}\left(m_{f+1}^2\right)$, one applies the matching condition, which takes the following form at the ℓ-loop level:

$$\alpha_{[f]}^{(\ell)}\left(m_{f+1}^2\right) = \sum_{j=1}^{\ell} k_j^{[f]}\left[\alpha_{[f+1]}^{(\ell)}\left(m_{f+1}^2\right)\right]^j, \tag{2.32}$$

where m_{f+1} stands for the $\overline{\text{MS}}$ mass of $(f+1)$th quark (see, e.g., papers [159, 185, 186] and references therein). Finally, the pair of values $\left[m_{f+1}^2, \alpha_{[f+1]}^{(\ell)}\left(m_{f+1}^2\right)\right]$ represents the boundary condition for the perturbative renormalization group equation at the ℓ-loop level (2.12), which enables one to evaluate the QCD invariant charge $\alpha_{[f+1]}^{(\ell)}(Q^2)$ at an arbitrary energy scale Q^2

within the kinematic interval $m_{f+1}^2 \leq Q^2 \leq m_{f+2}^2$. As in the one-loop case this can be performed in two alternative ways. Specifically, for this purpose one can either use the antiderivative of Eq. (2.12) or evaluate the ℓ-loop QCD scale parameter $\Lambda_{[f+1]}$ and employ the equivalent expression for the strong running coupling on hand (see, e.g., Eqs. 2.45, 2.55, 2.56, 2.57). Note that, similarly to Eq. (2.28), at the higher loop levels the matching condition (2.32) also allows one to directly relate the QCD scale parameters $\Lambda_{[f]}$ and $\Lambda_{[f+1]}$ to each other (see Eq. 2.41).

In Eq. (2.32) the one-loop coefficient k_1 has the native form

$$k_1 = 1, \tag{2.33}$$

which corresponds to the continuous matching condition (2.25). The next-order coefficient vanishes[6]

$$k_2 = 0, \tag{2.34}$$

which makes the two-loop condition (2.32) identical to the one-loop one (2.25). Starting from the three-loop level (i.e., for $j \geq 3$) the coefficients k_j assume nonzero values, that makes the matching condition (2.32) somewhat different from its one-loop form (2.25) and implies that the strong running coupling $\alpha_s(Q^2)$ becomes discontinuous at the quark mass thresholds. Specifically, the three-loop coefficient takes a quite simple form [187, 188]

$$k_3 = \frac{1}{\pi^2} \frac{11}{72}. \tag{2.35}$$

The higher-order ($j \geq 4$) coefficients k_j acquire dependence on the number of active flavors n_f. Namely, the four-loop coefficient reads [159, 186]

$$k_4^{[f]} = \frac{1}{\pi^3} \left[\frac{564731}{124416} - \frac{82043}{27648} \zeta(3) - \frac{2633}{31104} n_f \right], \tag{2.36}$$

whereas the five-loop coefficient takes the form [175–177]

$$
\begin{aligned}
k_5^{[f]} = \frac{1}{\pi^4} \Biggl\{ & \frac{291716893}{6123600} - \frac{2362581983}{87091200} \zeta(3) - \frac{76940219}{2177280} \zeta(4) \\
& + \frac{9318467}{362880} \zeta(4) \ln 2 - \frac{12057583}{483840} \zeta(5) + \frac{3031309}{435456} c_4 + \frac{340853}{90720} c_5 \\
& + n_f \left[-\frac{4770941}{2239488} + \frac{3645913}{995328} \zeta(3) - \frac{541549}{165888} \zeta(4) \right. \\
& \left. + \frac{115}{576} \zeta(5) + \frac{685}{41472} c_4 \right] + n_f^2 \left[-\frac{271883}{4478976} + \frac{167}{5184} \zeta(3) \right] \Biggr\}. \tag{2.37}
\end{aligned}
$$

---

6. Unless the matching scale is taken to be different from the respective quark mass threshold. In the latter case the matching condition for the strong running coupling differs from Eq. (2.32) beyond the one-loop level (see papers [175–177] and references therein).

TABLE 2.3 Numerical Values of the Coefficients $k_j^{[f]}$ $(1 \leq j \leq 5)$ Entering the Matching Condition for the Strong Running Coupling $\alpha_s(Q^2)$ at the $(f+1)$th Quark Mass Threshold (2.32)

$n_f = f$	k_1	k_2	$k_3 \times 10^2$	$k_4^{[f]} \times 10^2$	$k_5^{[f]} \times 10^2$
0	1.0000	0.0000	1.5480	3.1350	5.3079
1	1.0000	0.0000	1.5480	2.8620	4.2485
2	1.0000	0.0000	1.5480	2.5890	3.1440
3	1.0000	0.0000	1.5480	2.3160	1.9944
4	1.0000	0.0000	1.5480	2.0430	0.7997
5	1.0000	0.0000	1.5480	1.7700	−0.4402

In Eqs. (2.36), (2.37) $\zeta(x)$ denotes the Riemann ζ function (2.8),

$$c_4 = \frac{\ln^2 2}{3}\left(-\pi^2 + \ln^2 2\right) + 8\mathrm{Li}_4(1/2) \simeq 2.63615, \qquad (2.38)$$

$$c_5 = \frac{\ln^3 2}{3}\left(\frac{\pi^2}{3} - \frac{\ln^2 2}{5}\right) + 8\mathrm{Li}_5(1/2) \simeq 4.42174, \qquad (2.39)$$

and $\mathrm{Li}_t(x)$ stands for the polylogarithm function

$$\mathrm{Li}_t(x) = \sum_{n=1}^{\infty} \frac{x^n}{n^t}, \qquad (2.40)$$

$\mathrm{Li}_4(1/2) \simeq 0.51748$, $\mathrm{Li}_5(1/2) \simeq 0.50840$. The higher-order coefficients $k_j^{[f]}$ have not been calculated yet.

It is worthwhile to note that the matching condition for the strong running coupling at the $(f+1)$th quark mass threshold (2.32) involves the coefficients $k_j^{[f]}$ evaluated for $n_f = f$ active flavors. The numerical values of the coefficients $k_j^{[f]}$ at first five loop levels $(1 \leq j \leq 5)$ are listed in Table 2.3. As mentioned earlier, by making use of the matching condition (2.32) one can also directly relate the QCD scale parameters corresponding to $n_f = f$ and $n_f = f+1$ active flavors. In particular, at the five-loop level $(\ell = 5)$ such relation takes the form [177]

$$\beta_0^{[f]} \ln\left(\frac{\Lambda_{[f]}}{\Lambda_{[f+1]}}\right)^2 = \left(\beta_0^{[f]} - \beta_0^{[f+1]}\right)\lambda_{[f+1]} + \left(b_1^{[f]} - b_1^{[f+1]}\right)\ln\lambda_{[f+1]}$$

$$- b_1^{[f]}\ln\left(\frac{\beta_0^{[f]}}{\beta_0^{[f+1]}}\right) + \frac{1}{\beta_0^{[f+1]}\lambda_{[f+1]}}\left[b_1^{[f+1]}\left(b_1^{[f]} - b_1^{[f+1]}\right)\ln\lambda_{[f+1]}\right.$$

$$\left. + \left(b_1^{[f]}\right)^2 - \left(b_1^{[f+1]}\right)^2 - b_2^{[f]} + b_2^{[f+1]} + (4\pi)^2 k_3^{[f]}\right]$$

$$+ \frac{1}{\left(\beta_0^{[f+1]}\lambda_{[f+1]}\right)^2} \left\{ \frac{1}{2}\left(b_1^{[f+1]}\right)^3 \left(\ln^2 \lambda_{[f+1]} - 1\right) - b_1^{[f]}\left(b_1^{[f+1]}\right)^2 \right.$$

$$\times \left(\frac{1}{2}\ln^2 \lambda_{[f+1]} - \ln \lambda_{[f+1]} - 1\right) - b_1^{[f+1]}\left[\left(b_1^{[f]}\right)^2 - b_2^{[f]} + b_2^{[f+1]}\right.$$

$$\left. + (4\pi)^2 k_3^{[f]}\right]\ln \lambda_{[f+1]} - \frac{1}{2}\left(b_1^{[f]}\right)^3 + b_1^{[f]}\left[b_2^{[f]} - b_2^{[f+1]} - (4\pi)^2 k_3^{[f]}\right]$$

$$\left. -\frac{1}{2}\left(b_3^{[f]} - b_3^{[f+1]}\right) + (4\pi)^3 k_4^{[f]}\right\}$$

$$+ \frac{1}{\left(\beta_0^{[f+1]}\lambda_{[f+1]}\right)^3} \left\{ -\left(b_1^{[f+1]}\right)^4 \left(\frac{1}{3}\ln^3 \lambda_{[f+1]} - \frac{1}{2}\ln^2 \lambda_{[f+1]}\right.\right.$$

$$\left. - \ln \lambda_{[f+1]} - \frac{1}{6}\right) + b_1^{[f]}\left(b_1^{[f+1]}\right)^3 \left(\frac{1}{3}\ln^3 \lambda_{[f+1]} - \frac{3}{2}\ln^2 \lambda_{[f+1]}\right.$$

$$\left. - \ln \lambda_{[f+1]} + \frac{1}{2}\right) + \left(b_1^{[f+1]}\right)^2 \left[\left(b_1^{[f]}\right)^2 - b_2^{[f]} + b_2^{[f+1]}\right.$$

$$\left. + (4\pi)^2 k_3^{[f]}\right]\left(\ln^2 \lambda_{[f+1]} - \ln \lambda_{[f+1]} - 1\right) + b_1^{[f+1]}\left[\left(b_1^{[f]}\right)^3\right.$$

$$\left. -2b_1^{[f]}\left(b_2^{[f]} - b_2^{[f+1]} - (4\pi)^2 k_3^{[f]}\right) + b_3^{[f]} - b_3^{[f+1]} - 2(4\pi)^3 k_4^{[f]}\right]\ln \lambda_{[f+1]}$$

$$+ \frac{1}{3}\left(b_1^{[f]}\right)^4 - \left(b_1^{[f]}\right)^2 \left[b_2^{[f]} - b_2^{[f+1]} - (4\pi)^2 k_3^{[f]}\right] - (4\pi)^4 \left(k_3^{[f]}\right)^2$$

$$-\frac{1}{3}\left(b_4^{[f]} - b_4^{[f+1]}\right) + \frac{1}{3}\left(b_2^{[f]} - b_2^{[f+1]}\right)\left[b_2^{[f]} - 2b_2^{[f+1]} - 3(4\pi)^2 k_3^{[f]}\right]$$

$$\left. -\frac{1}{6}b_1^{[f+1]}b_3^{[f+1]} + b_1^{[f]}\left[\frac{2}{3}b_3^{[f]} - \frac{1}{2}b_3^{[f+1]} - (4\pi)^3 k_4^{[f]}\right] + (4\pi)^4 k_5^{[f]}\right\},$$

$$\tag{2.41}$$

where $b_j^{[f]} = \beta_j^{[f]}/\beta_0^{[f]}$ and $\lambda_{[f+1]} = \ln(m_{f+1}^2/\Lambda_{[f+1]}^2)$. The number of terms on the right-hand side of Eq. (2.41) is determined by the loop level on hand. Specifically, at the one-loop level it is enough to retain only the first term on the right-hand side of Eq. (2.41), which makes the latter identical to Eq. (2.28). In turn, at the two-loop level the first three terms on the right-hand side of Eq. (2.41) have to be retained, whereas at the three-loop and four-loop levels the terms proportional to, respectively, $(\beta_0^{[f+1]}\lambda_{[f+1]})^{-1}$ and $(\beta_0^{[f+1]}\lambda_{[f+1]})^{-2}$ have to be included. The detailed description of the matching procedure for the QCD running coupling can be found in papers [174–177] and references therein.

2.3 QCD INVARIANT CHARGE AT HIGHER LOOP LEVELS

Let us address now the strong running coupling $\alpha_s^{(\ell)}(Q^2)$ beyond the one-loop level. We begin by studying the two-loop ($\ell = 2$) perturbative renormalization group equation for the QCD invariant charge (2.12), which takes the following form:

$$\frac{d\ln\left[a_{\mathrm{s}}^{(2)}(\mu^2)\right]}{d\ln\mu^2} = -a_{\mathrm{s}}^{(2)}(\mu^2) - B_1\left[a_{\mathrm{s}}^{(2)}(\mu^2)\right]^2. \tag{2.42}$$

In this equation $a_{\mathrm{s}}^{(2)}(\mu^2) = \alpha_{\mathrm{s}}^{(2)}(\mu^2)\beta_0/(4\pi)$ stands for the two-loop perturbative QCD couplant and $B_1 = \beta_1/\beta_0^2$. Similarly to the one-loop case examined in the previous section, it is convenient to separate the variables in Eq. (2.42), that yields

$$-\frac{da_{\mathrm{s}}^{(2)}(\mu^2)}{\left[a_{\mathrm{s}}^{(2)}(\mu^2)\right]^2\left[1 + B_1 a_{\mathrm{s}}^{(2)}(\mu^2)\right]} = d\ln\mu^2. \tag{2.43}$$

Then integrating Eq. (2.43) in finite limits one eventually arrives at the transcendental equation, which determines the two-loop perturbative QCD running coupling:

$$\frac{1}{a_{\mathrm{s}}^{(2)}(Q^2)} - B_1\ln\left[1 + \frac{1}{B_1 a_{\mathrm{s}}^{(2)}(Q^2)}\right] = \ln\left(\frac{Q^2}{\Lambda^2}\right). \tag{2.44}$$

The QCD scale parameter Λ appearing on the right-hand side of this equation absorbs all the terms independent of the energy scale Q^2 and of the value of the strong running coupling at Q^2. Evidently, the two-loop parameter Λ entering Eq. (2.44) differs from that at the one-loop level (2.18), specifically

$$\Lambda^2 = Q_0^2 \exp\left\{-\frac{4\pi}{\beta_0}\frac{1}{\alpha_{\mathrm{s}}^{(2)}(Q_0^2)} + B_1\ln\left[1 + \frac{1}{B_1}\frac{4\pi}{\beta_0}\frac{1}{\alpha_{\mathrm{s}}^{(2)}(Q_0^2)}\right]\right\}. \tag{2.45}$$

It is worthwhile to note here that, in general, there exist other antiderivatives of the left-hand side of Eq. (2.43). For example, one may subtract from both sides of Eq. (2.44) a constant (such as $B_1\ln B_1$), which can be incorporated into the second term on the left-hand side of this equation and into the scale parameter Λ on its right-hand side (see also discussion of this issue in, e.g., Ref. [189]). It is also necessary to outline that, as mentioned in the previous section, the theoretical analysis of the experimental data for a physical observable, being performed at various loop levels, leads to the values of the QCD scale parameter Λ, which differ from each other. Additionally, at the higher loop levels, since the involved perturbative expansion coefficients acquire dependence on the applied subtraction scheme, the resulting value of the scale parameter Λ also becomes scheme-dependent.

It is also interesting to note that in late 1990s it was found [190–193] that the solution to Eq. (2.44) can be explicitly expressed in terms of the multivalued Lambert W function. The latter is defined as the function inverse to $x\exp x$, namely

$$W_k(x)\exp\left[W_k(x)\right] = x. \tag{2.46}$$

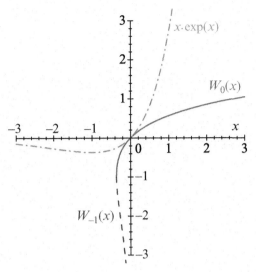

FIG. 2.5 The function xe^x (*dot-dashed curve*) and two real branches of the Lambert W function (2.46): the principal branch $W_0(x)$ (*solid curve*) and the branch $W_{-1}(x)$ (*dashed curve*).

This function, being known since the middle of 18th century, was applied to a wide range of topics in both mathematics and physics[7] (see review [199] for the details). The principal branch of the Lambert W function (2.46), $W_0(x)$, takes real values on the axis of real x for $-1/e \le x < \infty$, whereas the branch $W_{-1}(x)$ assumes real values on the axis of real x for $-1/e \le x < 0$, with $e \simeq 2.718$ being the base of natural logarithm (see Fig. 2.5). The other branches of the Lambert W function (2.46) acquire complex values. In the asymptotic $x \to \infty$, in the vicinity of $x = 0$, and at $x \to -\exp(-1)$ (in the latter case approaching the limit from the right being assumed) for the principal branch $W_0(x)$ the following expansions hold:

$$W_0(x) \simeq \ln x - \ln(\ln x) + \frac{\ln(\ln x)}{\ln x} + \frac{\ln^2(\ln x) - 2\ln(\ln x)}{2\ln^2 x} +$$
$$+ \frac{2\ln^3(\ln x) - 9\ln^2(\ln x) + 6\ln(\ln x)}{6\ln^3 x} + \mathcal{O}\left[\frac{\ln^4(\ln x)}{\ln^4 x}\right], \quad x \to \infty,$$

$$(2.47)$$

$$W_0(x) \simeq x - x^2 + \frac{3}{2}x^3 - \frac{8}{3}x^4 + \frac{125}{24}x^5 + \mathcal{O}\left(x^6\right), \quad x \to 0, \qquad (2.48)$$

---

7. For example, some applications of the Lambert W function within renormalization group method have been discussed in papers [194–198].

$$W_0\left(-\frac{1}{e}+x\right) \simeq -1+\sqrt{2ex} - \frac{1}{3}2ex + \frac{11}{72}(2ex)^{3/2} - \frac{43}{540}(2ex)^2 + \mathcal{O}\left(x^{5/2}\right),$$
$$x \to 0_+. \tag{2.49}$$

In turn, for $x \to -\exp(-1)$ (approaching the limit from the right being assumed) and for $x \to 0$ (approaching the limit from the left being assumed) the branch $W_{-1}(x)$ can be approximated by, respectively,

$$W_{-1}\left(-\frac{1}{e}+x\right) \simeq -1 - \sqrt{2ex} - \frac{1}{3}2ex - \frac{11}{72}(2ex)^{3/2}$$
$$- \frac{43}{540}(2ex)^2 + \mathcal{O}\left(x^{5/2}\right), \quad x \to 0_+, \tag{2.50}$$

and

$$W_{-1}(x) \simeq \ln(-x) - \ln(-\ln(-x)) + \frac{\ln(-\ln(-x))}{\ln(-x)}$$
$$+ \frac{\ln^2(-\ln(-x)) - 2\ln(-\ln(-x))}{2\ln^2(-x)}$$
$$+ \frac{2\ln^3(-\ln(-x)) - 9\ln^2(-\ln(-x)) + 6\ln(-\ln(-x))}{6\ln^3(-x)}$$
$$+ \mathcal{O}\left[\frac{\ln^4(-\ln(-x))}{\ln^4(-x)}\right], \quad x \to 0_-. \tag{2.51}$$

The detailed description of the properties of the Lambert W function (2.46) as well as a variety of its applications can be found in papers [199–202] and references therein.

To explicitly express the two-loop perturbative QCD running coupling $\alpha_s^{(2)}(Q^2)$ in terms of the Lambert W function one has to reduce the left-hand side of Eq. (2.44) to the form of the left-hand side of Eq. (2.46). For this purpose it is convenient to multiply both sides of Eq. (2.44) by $(-B_1^{-1})$ and subtract the unity from both sides of the obtained relation, that leads to

$$-\left[1+\frac{1}{B_1 a_s^{(2)}(Q^2)}\right] + \ln\left[1+\frac{1}{B_1 a_s^{(2)}(Q^2)}\right] = -\left(1+\frac{\ln z}{B_1}\right), \quad z = \frac{Q^2}{\Lambda^2}. \tag{2.52}$$

Then, one has to exponentiate this equation and multiply both sides of the resulting relation by (-1), which finally yields

$$-\left[1+\frac{1}{B_1 a_s^{(2)}(Q^2)}\right] \exp\left\{-\left[1+\frac{1}{B_1 a_s^{(2)}(Q^2)}\right]\right\} = -\exp\left[-\left(1+\frac{\ln z}{B_1}\right)\right]. \tag{2.53}$$

The form of the obtained equation is identical to that of Eq. (2.46), that, in turn, implies

$$-\left[1+\frac{1}{B_1 a_\mathrm{s}^{(2)}(Q^2)}\right]=W_k\left\{-\exp\left[-\left(1+\frac{\ln z}{B_1}\right)\right]\right\}. \qquad (2.54)$$

Since the QCD invariant charge $\alpha_\mathrm{s}(Q^2)$ is a real function of its argument, the branch index k on the right-hand side of Eq. (2.54) can take only two values, namely, either $k=0$ or $k=-1$. To make the choice between these two options, it is worthwhile to recall the physical features of the strong running coupling $\alpha_\mathrm{s}(Q^2)$. Specifically, it appears that the solution of Eq. (2.54), which corresponds to the principal branch of the Lambert W function ($k=0$), has to be discarded, because its asymptotic ultraviolet behavior (namely, $\alpha_\mathrm{s}^{(2)}(Q^2) \to -4\pi\beta_0/\beta_1$ when $Q^2 \to \infty$) is incompatible with QCD asymptotic freedom. At the same time the other solution of Eq. (2.54), which corresponds to the branch index $k=-1$, yields a physically sound behavior of the strong running coupling $\alpha_\mathrm{s}^{(2)}(Q^2)$ at high energies (specifically, $\alpha_\mathrm{s}(Q^2) \to 0_+$ when $Q^2 \to \infty$) and hence should be retained.

Thus the exact[8] solution of the two-loop renormalization group equation (2.42) for the QCD invariant charge can be represented in the following form (see also Refs. [190–193]):

$$\alpha_\mathrm{s}^{(2)}(Q^2)=-\frac{4\pi}{\beta_0}\frac{1}{B_1}\frac{1}{1+W_{-1}\left\{-\exp\left[-\left(1+B_1^{-1}\ln z\right)\right]\right\}}, \quad z=\frac{Q^2}{\Lambda^2},$$

$$(2.55)$$

where $B_1 = \beta_1/\beta_0^2$, the Lambert W function is defined in Eq. (2.46), and the scale parameter Λ is given by Eq. (2.45). It is worthwhile to mention here that, similar to the one-loop level studied in the previous section, the two-loop perturbative strong running coupling $\alpha_\mathrm{s}^{(2)}(Q^2)$ can also be expressed in terms of its value at a reference energy scale Q_0^2, the reference scale Q_0^2 itself, and the kinematic variable Q^2, specifically,

$$\alpha_\mathrm{s}^{(2)}(Q^2)=-\frac{4\pi}{\beta_0}\frac{1}{B_1}\frac{1}{1+W_{-1}\left[-A_0\exp\left(-A_0\right)\left(Q^2/Q_0^2\right)^{-1/B_1}\right]}, \qquad (2.56)$$

where

$$A_0=1+\frac{1}{B_1}\frac{4\pi}{\beta_0}\frac{1}{\alpha_\mathrm{s}^{(2)}(Q_0^2)}. \qquad (2.57)$$

Obviously, the two-loop perturbative QCD invariant charge $\alpha_\mathrm{s}^{(2)}(Q^2)$ given by the pair of expressions (2.45), (2.55) is equivalent to Eqs. (2.56), (2.57) and vice versa.

In general, the solution to the perturbative renormalization group equation for the strong running coupling $\alpha_\mathrm{s}^{(\ell)}(Q^2)$ (2.12) at any given loop level ℓ is

---

8. One has to bear in mind that Eq. (2.55) is the "exact" solution of the two-loop perturbative renormalization group equation (2.42), which is itself approximate.

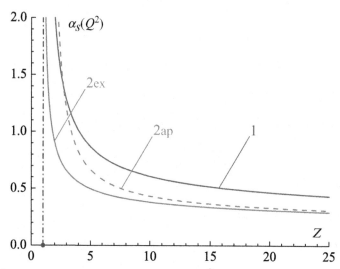

FIG. 2.6 The perturbative QCD invariant charge $\alpha_s(Q^2)$ at the one-loop level (Eq. 2.19, *solid curve*, label "1") and at the two-loop level: exact expression (Eq. 2.55, *solid curve*, label "2ex") and its approximate form (Eq. 2.58, *dashed curve*, label "2ap"). The values of parameters: $n_f = 3$ active flavors, $z = Q^2/\Lambda^2$. The location of the unphysical "ghost pole" $(Q^2 = \Lambda^2)$ is marked by the *vertical dot-dashed line*.

determined up to the terms of the order of $\mathcal{O}\left[\ln^\ell(\ln z)/\ln^{\ell+1} z\right]$, which correspond to the neglected higher-order contributions to the perturbative approximation of the β function appearing on the right-hand side of Eq. (2.12). In turn, this fact allows one to use in practical applications an approximate expression for the QCD invariant charge at the higher loop levels. In particular, by making use of expansion (2.51) one can demonstrate that in the ultraviolet asymptotic the two-loop strong running coupling $\alpha_s^{(2)}(Q^2)$ (2.55) takes the form

$$\alpha_s^{(2)}(Q^2) \simeq \frac{4\pi}{\beta_0} \frac{1}{\ln z + B_1 \ln\left(1 + B_1^{-1}\ln z\right)}$$

$$\simeq \frac{4\pi}{\beta_0}\left[\frac{1}{\ln z} - B_1\frac{\ln(\ln z)}{\ln^2 z}\right], \quad Q^2 \to \infty, \qquad (2.58)$$

the terms of the order of $\mathcal{O}\left[\ln^2(\ln z)/\ln^3 z\right]$ being neglected. The discussion of this issue will be given below and in Appendix A.

Fig. 2.6 displays the exact expression for the two-loop perturbative QCD invariant charge (2.55) along with its approximate form (2.58). As one can infer from this figure, at high energies these two expressions are indistinguishable from each other (within given accuracy), whereas at low energies their behavior turns out to be quantitatively different. At the same time the appreciable deviation of the expressions (2.55), (2.58) from each other marks the boundary

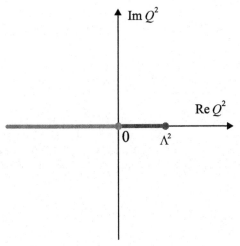

FIG. 2.7 The singularities of the two-loop perturbative strong running coupling $\alpha_s^{(2)}(Q^2)$ (2.55) in the complex Q^2-plane: cut $Q^2 \leq 0$ along the negative semiaxis of real Q^2 and unphysical cut $0 < Q^2 \leq \Lambda^2$ along the positive semiaxis of real Q^2.

of applicability range of the two-loop approximation (2.42) and indicates that the higher-loop corrections become nonnegligible.

Fig. 2.6 also implies that the exact two-loop perturbative strong running coupling (2.55) as well as its approximate form (2.58) contain unphysical singularities in the infrared domain. In particular, by making use of the expansion (2.50) it is straightforward to demonstrate that the QCD invariant charge $\alpha_s^{(2)}(Q^2)$ (2.55) diverges at low energies, namely

$$\alpha_s^{(2)}(Q^2) \simeq \frac{4\pi}{\beta_0}\sqrt{\frac{1}{2B_1}\frac{\Lambda^2}{Q^2 - \Lambda^2}}, \quad Q^2 \to \Lambda^2. \tag{2.59}$$

In this equation it is assumed that Q^2 approaches Λ^2 from the right. The detailed analysis of the analytic properties of the two-loop QCD invariant charge $\alpha_s^{(2)}(Q^2)$ (2.55) in the kinematic variable Q^2 reveals that the incorporation of the higher loop corrections does not eliminate the unphysical singularities in the solution to the perturbative renormalization group equation (2.12). On the contrary, at the two-loop level the perturbative QCD running coupling $\alpha_s^{(2)}(Q^2)$ (2.55) acquires an additional[9] unphysical singularity in Q^2 variable. Specifically, in the complex Q^2-plane the function $\alpha_s^{(2)}(Q^2)$ (2.55) possesses the cut $Q^2 \leq 0$ along the negative semiaxis of real Q^2, which can be called physical if all the involved particles are assumed to be massless, and the unphysical cut $0 < Q^2 \leq \Lambda^2$ along the positive semiaxis of real Q^2 (see Figs. 2.3 and 2.7 for the details).

---

9. With respect to the one-loop perturbative strong running coupling (2.19).

Let us proceed now to the three-loop level ($\ell = 3$). In this case the renormalization group equation for the QCD invariant charge (2.12) takes the following form

$$\frac{d\ln\left[a_s^{(3)}(\mu^2)\right]}{d\ln\mu^2} = -a_s^{(3)}(\mu^2) - B_1\left[a_s^{(3)}(\mu^2)\right]^2 - B_2\left[a_s^{(3)}(\mu^2)\right]^3, \quad (2.60)$$

where $B_1 = \beta_1/\beta_0^2$ and $B_2 = \beta_2/\beta_0^3$. Similar to the one-loop and two-loop cases examined earlier, one may separate the variables and integrate the resulting relation in finite limits that eventually leads to the following transcendental equation, which determines the three-loop strong running coupling:

$$\frac{1}{a_s^{(3)}(Q^2)} + B_1\ln\left\{\frac{a_s^{(3)}(Q^2)}{\sqrt{1 + B_1 a_s^{(3)}(Q^2) + B_2\left[a_s^{(3)}(Q^2)\right]^2}}\right\}$$

$$+ \frac{B_1^2 - 2B_2}{\sqrt{B_1^2 - 4B_2}}\tanh^{-1}\left[\frac{B_1 + 2B_2 a_s^{(3)}(Q^2)}{\sqrt{B_1^2 - 4B_2}}\right] = \ln\left(\frac{Q^2}{\Lambda^2}\right). \quad (2.61)$$

In this equation $\tanh^{-1}(x)$ stands for the inverse hyperbolic tangent of x and is not to be confused with the multiplicative inverse $1/\tanh(x)$. As earlier, the QCD scale parameter Λ entering[10] Eq. (2.61) incorporates all the terms independent of the energy scale Q^2 and of the value of the strong running coupling at Q^2. However, it turns out that Eq. (2.61) cannot be explicitly solved in terms of the presently known functions. Nonetheless, at the higher loop levels (i.e., for $\ell \geq 2$) the perturbative renormalization group equation (2.12) is commonly solved by making use of the iterative method, which provides the expression for the QCD invariant charge $\alpha_s^{(\ell)}(Q^2)$ indistinguishable from the exact solution to Eq. (2.12) within given accuracy (see the discussion of this issue below and in Appendix A).

At the same time it is interesting to mention that the expression for the strong running coupling, which differs from the exact solution to Eq. (2.60) by the corrections uncontrollable at the three-loop level, can also be explicitly represented in terms of the aforementioned Lambert W function (2.46) by making use of the mathematical trick discussed in paper [191]. Specifically, the three-loop perturbative β function appearing on the right-hand side of Eq. (2.60) can be substituted by the so-called Padé approximant of the order $(2, 1)$

$$\beta_{\mathrm{P21}}^{(3)}(a_s) = -\frac{a_s + a_s^2(B_1 - B_2/B_1)}{1 - a_s B_2/B_1}. \quad (2.62)$$

---

10. Note that the three-loop QCD scale parameter Λ appearing on the right-hand side of Eq. (2.61) differs from its one-loop and two-loop expressions (see Eqs. 2.18, 2.45, respectively).

Certainly, such a replacement is valid in the ultraviolet asymptotic only. In particular, for the small values of the strong running coupling the expression (2.62) coincides with the three-loop perturbative β function (2.2) up to the terms of the order of $\mathcal{O}\left(a_s^4\right)$, which are uncontrollable at the loop level on hand:

$$\beta_{\text{P21}}^{(3)}(a_s) \simeq -a_s - B_1 a_s^2 - B_2 a_s^3 + \mathcal{O}\left(a_s^4\right) = \beta_{\text{pert}}^{(3)}(a_s) + \mathcal{O}\left(a_s^4\right), \quad a_s \to 0_+. \tag{2.63}$$

However, it is necessary to outline that, in general, the Padé approximation of the β function perturbative expansion substantially affects the infrared behavior of the resulting QCD invariant charge. Specifically, the solution of the ℓ-loop renormalization group equation (2.12) at low energies can be represented as

$$\alpha_s^{(\ell)}(Q^2) \simeq \frac{4\pi}{\beta_0} \left(\frac{1}{\ell B_{\ell-1}} \frac{\Lambda^2}{Q^2 - \Lambda^2} \right)^{1/\ell}, \quad Q^2 \to \Lambda^2 \tag{2.64}$$

(with $B_{\ell-1} > 0$ being assumed), whereas the approximation (2.62) eventually results in the expression

$$\alpha_{\text{P21}}^{(3)}(Q^2) \simeq \frac{4\pi}{\beta_0} \frac{B_2}{B_2 - B_1^2} \frac{\Lambda^2}{Q^2 - \Lambda^2}, \quad Q^2 \to \Lambda^2, \tag{2.65}$$

which evidently differs from Eq. (2.64) for $\ell = 3$.

Thus the renormalization group equation for the strong running coupling $\alpha_{\text{P21}}^{(3)}(Q^2)$, which corresponds to the approximate three-loop β function (2.62), acquires the following form

$$\frac{d \ln\left[a_{\text{P21}}^{(3)}(\mu^2)\right]}{d \ln \mu^2} = -a_{\text{P21}}^{(3)}(\mu^2) \frac{1 + a_{\text{P21}}^{(3)}(\mu^2)(B_1 - B_2/B_1)}{1 - a_{\text{P21}}^{(3)}(\mu^2)B_2/B_1}. \tag{2.66}$$

Once again, separating the variables and integrating the obtained equation in finite limits, one arrives at

$$\frac{1}{a_{\text{P21}}^{(3)}(Q^2)} - B_1 \ln\left[\frac{1}{B_1 a_{\text{P21}}^{(3)}(Q^2)} + 1 - \frac{B_2}{B_1^2} \right] = \ln\left(\frac{Q^2}{\Lambda^2}\right). \tag{2.67}$$

As earlier, the scale parameter Λ entering this equation absorbs all the terms, which do not depend on the kinematic variable Q^2 and on the value of the QCD invariant charge at Q^2. Following the very same lines as in the two-loop case, one can cast the left-hand side of Eq. (2.67) into the form identical to the left-hand side of Eq. (2.46), specifically

$$\left[\frac{B_2}{B_1^2} - 1 - \frac{1}{B_1 a_{\text{P21}}^{(3)}(Q^2)} \right] \exp\left[\frac{B_2}{B_1^2} - 1 - \frac{1}{B_1 a_{\text{P21}}^{(3)}(Q^2)} \right]$$

$$= -\exp\left(\frac{B_2}{B_1^2} - 1 - \frac{\ln z}{B_1} \right), \tag{2.68}$$

where $z = Q^2/\Lambda^2$. In turn, Eqs. (2.46), (2.68) imply that

$$\frac{B_2}{B_1^2} - 1 - \frac{1}{B_1 a_{\mathrm{P21}}^{(3)}(Q^2)} = W_k\left[-\exp\left(\frac{B_2}{B_1^2} - 1 - \frac{\ln z}{B_1}\right)\right]. \tag{2.69}$$

Similarly to the aforementioned two-loop level, the fact that the QCD invariant charge is a real function of its argument forces one to discard all the values of the branch index k in Eq. (2.69) except for $k = 0$ and $k = -1$. But only the latter one yields a physically justified behavior of the strong running coupling at high energies (namely, $\alpha_{\mathrm{P21}}^{(3)}(Q^2) \to 0_+$ when $Q^2 \to \infty$), which meets the requirement of the QCD asymptotic freedom and therefore should be retained. Thus, the solution of the renormalization group equation for the QCD invariant charge, which corresponds to the Padé approximant of the order of $(2, 1)$ of the three-loop perturbative β function (2.62), takes the following form (see also paper [191])

$$\alpha_{\mathrm{P21}}^{(3)}(Q^2) = -\frac{4\pi}{\beta_0} \frac{1}{B_1} \frac{1}{1 + W_{-1}\left\{-\exp\left[-\left(1 + B_1^{-1}\ln z - B_2/B_1^2\right)\right]\right\} - B_2/B_1^2}. \tag{2.70}$$

In this equation $z = Q^2/\Lambda^2$, $B_1 = \beta_1/\beta_0^2$, $B_2 = \beta_2/\beta_0^3$, and the Lambert W function is defined in Eq. (2.46). Obviously, for the case of $\beta_2 = 0$ the Padé approximant for the three-loop β function (2.62) acquires the form of the two-loop perturbative expression $\beta_{\mathrm{pert}}^{(2)}(a_s) = -a_s - B_1 a_s^2$, and the strong running coupling (2.70) becomes identical to Eq. (2.55). As earlier, an approximate form of Eq. (2.70), which is applicable in the ultraviolet asymptotic $Q^2 \to \infty$, can be obtained by making use of the expansion (2.51), specifically

$$\alpha_{\mathrm{P21}}^{(3)}(Q^2) \simeq \frac{4\pi}{\beta_0} \frac{1}{\ln z + B_1 \ln\left(1 - B_1^{-2}B_2 + B_1^{-1}\ln z\right)\left[1 + \left(1 - B_1^{-2}B_2 + B_1^{-1}\ln z\right)^{-1}\right]}$$
$$\simeq \frac{4\pi}{\beta_0}\left\{\frac{1}{\ln z} - B_1\frac{\ln(\ln z)}{\ln^2 z} + \frac{1}{\ln^3 z}\left[B_1^2\left(\ln^2(\ln z) - \ln(\ln z) - 1\right) + B_2\right]\right\}, \tag{2.71}$$

the terms of the order of $\mathcal{O}\left[\ln^3(\ln z)/\ln^4 z\right]$ being neglected (see the discussion of this issue below and in Appendix A).

As mentioned earlier in practical applications one commonly employs the iterative solution to the perturbative renormalization group equation for the QCD invariant charge $\alpha_s(Q^2)$, which approximates its exact form accurately enough. As an example, it is worthwhile to derive such solution at the three-loop level ($\ell = 3$). For this purpose it is convenient to separate the variables in Eq. (2.12), which casts it to

$$-\frac{da_s}{a_s^2\left(1 + B_1 a_s + B_2 a_s^2\right)} = d\ln\mu^2, \tag{2.72}$$

where $a_s(\mu^2) = \alpha_s(\mu^2)\beta_0/(4\pi)$ stands for the QCD couplant and $B_j = \beta_j/\beta_0^{j+1}$ is the ratio of the β function perturbative expansion coefficients. Then because for the small values of the QCD couplant

$$\frac{1}{1 + B_1 a_s + B_2 a_s^2} \simeq 1 - B_1 a_s - (B_2 - B_1^2)a_s^2 + \mathcal{O}\left(a_s^3\right), \tag{2.73}$$

within the accuracy of the three-loop level Eq. (2.72) can be represented as

$$\frac{da_s}{a_s^2}\left[-1 + B_1 a_s + (B_2 - B_1^2)a_s^2\right] = d\ln\mu^2. \tag{2.74}$$

In turn, the integration of this equation in finite limits results in

$$\frac{1}{a_s(Q^2)} + B_1 \ln\left[a_s(Q^2)\right] + (B_2 - B_1^2)a_s(Q^2) = \ln z, \quad z = \frac{Q^2}{\Lambda^2}. \tag{2.75}$$

As earlier the QCD scale parameter Λ appearing on the right-hand side of this equation incorporates all the terms independent of the energy scale Q^2 and of the value of the QCD couplant at Q^2.

Obviously, for the small values of $a_s(Q^2)$ the first term on the left-hand side of Eq. (2.75) is the leading one, whereas the second and third terms are subleading and subsubleading ones, respectively. Therefore the first iteration of Eq. (2.75) yields

$$a_s^{(1)}(Q^2) = \frac{1}{\ln z}, \tag{2.76}$$

which coincides with the exact one-loop result (2.19). Then the second iteration of Eq. (2.75) leads to

$$\frac{1}{a_s^{(2)}(Q^2)} = \ln z - B_1 \ln\left[a_s^{(1)}(Q^2)\right] = \ln z + B_1 \ln(\ln z). \tag{2.77}$$

Hence, in the ultraviolet asymptotic $z \to \infty$

$$a_s^{(2)}(Q^2) = \frac{1}{\ln z} - B_1 \frac{\ln(\ln z)}{\ln^2 z}, \tag{2.78}$$

which coincides with the approximate two-loop solution obtained earlier (see Eq. 2.58). And finally the third iteration of Eq. (2.75) gives

$$\frac{1}{a_s^{(3)}(Q^2)} = \ln z - B_1 \ln\left[a_s^{(2)}(Q^2)\right] - (B_2 - B_1^2)a_s^{(2)}(Q^2)$$

$$\simeq \ln z + B_1 \ln(\ln z) + \frac{1}{\ln z}\left[B_1^2 \ln(\ln z) + B_1^2 - B_2\right], \quad z \to \infty. \tag{2.79}$$

Thus the iterative solution of the perturbative renormalization group equation for the strong running coupling at the three-loop level ($\ell = 3$) takes the form

$$\alpha_s^{(3)}(Q^2) = \frac{4\pi}{\beta_0} \left\{ \frac{1}{\ln z} - B_1 \frac{\ln(\ln z)}{\ln^2 z} + \frac{1}{\ln^3 z} \left[B_1^2 \left(\ln^2(\ln z) - \ln(\ln z) - 1 \right) + B_2 \right] \right\},$$

(2.80)

which coincides with Eq. (2.71). In Eq. (2.80) $z = Q^2/\Lambda^2$, $B_j = \beta_j/\beta_0^{j+1}$, and $Q^2 \to \infty$ is assumed.

Following the very same lines as above it is straightforward to demonstrate that the iterative solution to Eq. (2.12) at the four-loop and five-loop levels in the ultraviolet asymptotic can be represented as, respectively,

$$\alpha_s^{(4)}(Q^2) = \frac{4\pi}{\beta_0} \left\{ \frac{1}{\ln z} - B_1 \frac{\ln(\ln z)}{\ln^2 z} + \frac{1}{\ln^3 z} \left[B_1^2 \left(\ln^2(\ln z) - \ln(\ln z) - 1 \right) + B_2 \right] \right.$$
$$+ \frac{1}{\ln^4 z} \left[B_1^3 \left(-\ln^3(\ln z) + \frac{5}{2} \ln^2(\ln z) + 2\ln(\ln z) - \frac{1}{2} \right) \right.$$
$$\left. \left. -3B_1 B_2 \ln(\ln z) + \frac{B_3}{2} \right] \right\},$$

(2.81)

and

$$\alpha_s^{(5)}(Q^2) = \frac{4\pi}{\beta_0} \left\{ \frac{1}{\ln z} - B_1 \frac{\ln(\ln z)}{\ln^2 z} + \frac{1}{\ln^3 z} \left[B_1^2 \left(\ln^2(\ln z) - \ln(\ln z) - 1 \right) + B_2 \right] \right.$$
$$+ \frac{1}{\ln^4 z} \left[B_1^3 \left(-\ln^3(\ln z) + \frac{5}{2} \ln^2(\ln z) + 2\ln(\ln z) - \frac{1}{2} \right) \right.$$
$$\left. -3B_1 B_2 \ln(\ln z) + \frac{B_3}{2} \right] + \frac{1}{\ln^5 z} \left[B_1^4 \left(\ln^4 \ln z - \frac{13}{3} \ln^3 \ln z \right. \right.$$
$$\left. - \frac{3}{2} \ln^2 \ln z + 4 \ln \ln z + \frac{7}{6} \right) + 3B_1^2 B_2 \left(2\ln^2 \ln z - \ln \ln z - 1 \right)$$
$$\left. \left. -B_1 B_3 \left(2\ln \ln z + \frac{1}{6} \right) + \frac{5}{3} B_2^2 + \frac{B_4}{3} \right] \right\}.$$

(2.82)

It is also worthwhile to mention that at the higher loop levels (i.e., for $\ell \geq 2$) the terms that constitute the difference between the exact solution to the perturbative renormalization group equation for the QCD invariant charge (2.12) and the iterative one are beyond the accuracy of the loop level on hand.

Fig. 2.8 displays the perturbative expressions for the strong running coupling at the first five loop levels, indicated by numerical labels (see Eqs. 2.19, 2.78, 2.80–2.82). As one can infer from this figure at high energies all these functions tend to each other and the deviation between curves corresponding to the ℓ-loop and ($\ell + 1$)-loop levels becomes less pronounced with increasing ℓ. However, at low energies all the expressions for the perturbative QCD invariant charge $\alpha_s(Q^2)$ diverge due to the presence of unphysical singularities.

As one may note at an arbitrary loop level the strong running coupling obtained as the iterative solution to the perturbative renormalization group equation (2.12) can be represented as the double sum

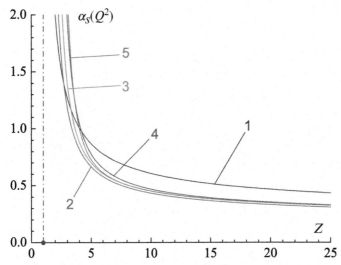

FIG. 2.8 The perturbative strong running coupling $\alpha_s(Q^2)$ at the first five loop levels: one-loop expression (2.19) and higher-loop iterative solutions (2.78), (2.80)–(2.82). The loop level is indicated by the *numerical labels*. The values of parameters: $n_f = 3$ active flavors, $z = Q^2/\Lambda^2$. The location of the unphysical "ghost pole" $(Q^2 = \Lambda^2)$ is marked by the *vertical dot-dashed line*.

$$\alpha_s^{(\ell)}(Q^2) = \frac{4\pi}{\beta_0} \sum_{n=1}^{\ell} \sum_{m=0}^{n-1} b_n^m \frac{\ln^m(\ln z)}{\ln^n z}, \quad z = \frac{Q^2}{\Lambda^2}, \quad Q^2 \to \infty, \tag{2.83}$$

where b_n^m (the integer superscript m is not to be confused with respective power) denotes the combination of the β function perturbative expansion coefficients. At the first few loop levels the coefficients b_n^m assume quite simple form (e.g., $b_1^0 = 1, b_2^0 = 0, b_2^1 = -B_1, b_3^0 = B_2 - B_1^2, b_3^1 = -B_1^2, b_3^2 = B_1^2$, etc.), whereas at the higher loop levels the explicit expressions for b_n^m become rather cumbersome (see Appendix A for the details). It is also necessary to outline that at any given loop level the exact solution to the perturbative renormalization group equation for the QCD invariant charge, as well as the iterative one, possesses unphysical singularities in the kinematic variable Q^2 at low energies. As mentioned earlier the presence of such singularities contradicts the basic principles of the theory and makes the perturbative approach inapplicable to the study of the strong interaction processes in the infrared domain, which constitutes its inherent obstacle.

Chapter 3

Functions $\Pi(q^2)$, $R(s)$, and $D(Q^2)$ Within Perturbative Approach

The asymptotic freedom of Quantum Chromodynamics (QCD) enables one to study a variety of strong interaction processes at high energies in the framework of perturbation theory. Despite the fact that the latter inevitably entails such difficulties as the infrared unphysical singularities and the renormalization-scheme dependence of the higher loop corrections, the QCD perturbative calculations constitute the basis of theoretical description of hadron dynamics in the ultraviolet domain and still remain the basic tool commonly employed in practical applications.

3.1 HADRONIC VACUUM POLARIZATION FUNCTION WITHIN PERTURBATIVE APPROACH

Let us start with the hadronic vacuum polarization function $\Pi(q^2)$ (1.28) calculated in the leading order of perturbation theory, namely, in the first order in the electromagnetic coupling and in the zeroth order in the strong coupling. The pertinent contribution to the photon self-energy is given by the quark-antiquark loop displayed in Fig. 3.1 and the corresponding hadronic vacuum polarization tensor $\Pi_{\mu\nu}(q^2)$ takes the form

$$\Pi_{\mu\nu}^{(0)}(q^2) = -N_c \sum_{f=1}^{n_f} Q_f^2 \int \frac{d^4k}{(2\pi)^4} \mathrm{Tr} \left\{ \gamma_\mu \frac{1}{\hat{q} + \hat{k} - m_f + i0_+} \gamma_\nu \frac{1}{\hat{k} - m_f + i0_+} \right\}.$$

$$(3.1)$$

In this equation $N_c = 3$ is the number of colors, n_f denotes the number of active flavors, Q_f stands for the electric charge of the quark of fth flavor (in units of the elementary charge e), m_f is the mass of the quark of fth flavor, and the notation $\hat{k} = \gamma^\mu k_\mu$ is used. On the right-hand side of Eq. (3.1) the factor (-1) appears due to the fermion loop. In quark propagators the infinitesimal imaginary term will be omitted hereinafter.

Strong Interactions in Spacelike and Timelike Domains. http://dx.doi.org/10.1016/B978-0-12-803439-2.00003-X

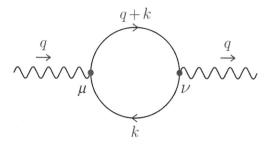

FIG. 3.1 The leading order perturbative contribution (first order in the electromagnetic coupling and zeroth order in the strong coupling) to the hadronic vacuum polarization function $\Pi(q^2)$ (1.28).

First of all, it is convenient to rewrite Eq. (3.1) in the following way:

$$\Pi_{\mu\nu}^{(0)}(q^2) = -N_c \sum_{f=1}^{n_f} Q_f^2 \int \frac{d^4k}{(2\pi)^4} \frac{\text{Tr}\left[\gamma_\mu(\hat{k} + \hat{q} + m_f)\gamma_\nu(\hat{k} + m_f)\right]}{\left[(k+q)^2 - m_f^2\right]\left[k^2 - m_f^2\right]}. \tag{3.2}$$

To calculate the hadronic vacuum polarization tensor (3.2) the Feynman parameterization will be employed. This method is based on the identical transformation of the integrand of Eq. (3.2) to the form, which allows one to easily perform the integration over the relevant kinematic variable. Specifically, the product of n arbitrary terms $1/v_j$ can be represented as

$$\frac{1}{v_1 v_2 \ldots v_n} = (n-1)! \int_0^1 dx_1 \int_0^1 dx_2 \cdots \int_0^1 dx_n \left(\sum_{j-1}^n v_j x_j\right)^{-n}, \tag{3.3}$$

where

$$\sum_{j=1}^n x_j = 1. \tag{3.4}$$

The constraint (3.4) on the variables x_j eliminates one integration on the right-hand side of Eq. (3.3). For example, for $n = 2$ Eq. (3.3) takes the form

$$\frac{1}{v_1 v_2} = \int_0^1 \frac{dx}{[v_1 x + v_2(1-x)]^2}. \tag{3.5}$$

In the considered case v_1 and v_2 are the functions of quark's momenta squared and mass squared:

$$\frac{1}{\left[(k+q)^2 - m_f^2\right]\left[k^2 - m_f^2\right]} = \int_0^1 \frac{dx}{\left\{\left[(k+q)^2 - m_f^2\right]x + \left[k^2 - m_f^2\right](1-x)\right\}^2}. \tag{3.6}$$

Hence, the hadronic vacuum polarization tensor (3.2) can be equivalently rewritten as

$$\Pi^{(0)}_{\mu\nu}(q^2) = -N_c \sum_{f=1}^{n_f} Q_f^2 \int_0^1 dx \int \frac{d^4k}{(2\pi)^4} \frac{\text{Tr}\left\{\gamma_\mu \left[\hat{k} + \hat{q} + m_f\right] \gamma_\nu \left[\hat{k} + m_f\right]\right\}}{\left[(k + qx)^2 + q^2 x(1 - x) - m_f^2\right]^2}.$$

(3.7)

In what follows the limit of the massless quarks ($m_f = 0$) will be assumed. As discussed in Section 1.1, in this case the only difference between the leptonic and quark contributions to the photon self-energy consists in the prefactor $N_c \sum_{f=1}^{n_f} Q_f^2$ appearing on the right-hand side of Eq. (3.1). In Eq. (3.7) it is convenient to change the integration variable $k_\mu = p_\mu - q_\mu x$, which yields

$$\Pi^{(0)}_{\mu\nu}(q^2) = -N_c \sum_{f=1}^{n_f} Q_f^2 \int_0^1 dx \int \frac{d^4p}{(2\pi)^4} \frac{\text{Tr}\left\{\gamma_\mu \left[\hat{p} + \hat{q}(1 - x)\right] \gamma_\nu \left[\hat{p} - \hat{q}x\right]\right\}}{\left[p^2 + q^2 x(1 - x)\right]^2}.$$

(3.8)

The numerator of the integrand of this equation can be calculated by making use of Eq. (1.8), specifically

$$\text{Tr}\left\{\gamma_\mu \left[\hat{p} + \hat{q}(1 - x)\right] \gamma_\nu \left[\hat{p} - \hat{q}x\right]\right\} = 4\left\{\left(g_{\mu\nu} q^2 - 2q_\mu q_\nu\right) x(1 - x)\right.$$
$$\left. + 2p_\mu p_\nu - g_{\mu\nu} p^2 + \left[p_\mu q_\nu + q_\mu p_\nu - g_{\mu\nu}(pq)\right](1 - 2x)\right\}.$$

(3.9)

It is worth noting here that the calculation of the hadronic vacuum polarization tensor (3.7) in the case $m_f \neq 0$ also involves Eqs. (1.6), (1.7).

Since Eq. (3.9) constitutes a part of the integrand of Eq. (3.8), it can be simplified by virtue of the pertinent symmetry constraints. In particular, the integral of the product of a function $f(p^2)$ and odd number of vectors p_μ over d^4p vanishes:

$$\int d^4p \, f(p^2) \prod_{j=1}^{2n+1} p_{\mu_j} = 0,$$

(3.10)

with $n = 0, 1, 2, 3, \ldots$ being a nonnegative integer. In turn, this allows one to discard the terms proportional to $(1 - 2x)$ in the second line of Eq. (3.9) in further calculations. Additionally, the integral of the product of a function $f(p^2)$ and even number of vectors p_μ over d^4p can be simplified in the following way:

$$\int d^4p \, f(p^2) p_\mu p_\nu = \frac{1}{4} g_{\mu\nu} \int d^4p \, f(p^2) p^2.$$

(3.11)

In turn, these relations allow one to combine two first terms in the second line of Eq. (3.9) into a single one. Therefore, the hadronic vacuum polarization tensor (3.8) can be represented as

$$\Pi_{\mu\nu}^{(0)}(q^2) = 4N_c \sum_{f=1}^{n_f} Q_f^2 \int_0^1 dx \int \frac{d^4p}{(2\pi)^4} \frac{g_{\mu\nu}(p^2/2) + \left(2q_\mu q_\nu - q^2 g_{\mu\nu}\right) x(1-x)}{\left[p^2 + q^2 x(1-x)\right]^2}.$$

(3.12)

Then one can perform Wick rotation

$$p^0 \to i p_E^0, \quad \vec{p} \to \vec{p}_E, \quad p^2 \to -p_E^2, \quad \int d^4p \to i \int d^4p_E$$

(3.13)

to transform the expression on hand to the four-dimensional Euclidean momentum space, namely (the subscript "E" will be omitted hereinafter)

$$\Pi_{\mu\nu}^{(0)}(q^2) = -4iN_c \sum_{f=1}^{n_f} Q_f^2 \int_0^1 dx \int \frac{d^4p}{(2\pi)^4} \frac{g_{\mu\nu}\left(p^2/2\right) - \left(2q_\mu q_\nu - q^2 g_{\mu\nu}\right) x(1-x)}{\left[p^2 - q^2 x(1-x)\right]^2}.$$

(3.14)

As one can infer from Eq. (3.14) the four-dimensional momentum integral diverges. To isolate the divergent part in Eq. (3.14) a method of regularization has to be applied. It is worthwhile to mention here that, in general, there is no physically motivated rigorous criteria that makes one method of regularization more favorable than the others. In what follows the method of dimensional regularization [203–206] will be employed (see also reviews [142–146]). This method is based on the continuation of the integral over the four-dimensional momentum space d^4p to the integral over momentum space of the noninteger dimension $n = 4 - 2\varepsilon$. At the same time one has to be aware that such a procedure entails certain difficulties related to the definition of the metric and the complete set of γ matrices (in particular, the matrix γ_5) in the space of a noninteger dimension (see Refs. [150, 207] for a discussion of this issue).

Thus the regularized hadronic vacuum polarization tensor (3.14) acquires the following form:

$$\text{reg}\,\Pi_{\mu\nu}^{(0)}(q^2) = -4iN_c \sum_{f=1}^{n_f} Q_f^2 \int_0^1 dx \mu^{4-n} \int \frac{d^n p}{(2\pi)^n} \frac{1}{\left[p^2 - q^2 x(1-x)\right]^2}$$

$$\times \left[g_{\mu\nu} \frac{p^2}{2} - \left(2q_\mu q_\nu - q^2 g_{\mu\nu}\right) x(1-x)\right].$$

(3.15)

Note that the parameter μ of the dimension of mass has been introduced into the expression (3.15) to keep its dimension unchanged. Then in Eq. (3.15) the integration over $d^n p$ can be performed by making use of the master integral

$$\int d^n p \frac{(p^2)^a}{(p^2 + M^2)^b} = \pi^{n/2} \frac{\Gamma(b-a-n/2)\Gamma(a+n/2)}{\Gamma(n/2)\Gamma(b)} \left(M^2\right)^{a-b+n/2},$$

(3.16)

with $\Gamma(y)$ being the Euler Γ function

$$\Gamma(y) = \int\limits_0^{\infty} e^{-t} t^{y-1} dt, \quad \Gamma(y+1) = y\Gamma(y). \tag{3.17}$$

Specifically, the integral over $d^n p$ of the first term in the square brackets in the second line of Eq. (3.15) corresponds to the parameters $a = 1$, $b = 2$, and $M^2 = -q^2 x(1-x)$ in Eq. (3.16), that leads to

$$\int \frac{p^2 d^n p}{\left[p^2 - q^2 x(1-x)\right]^2} = \pi^{n/2} \frac{\Gamma(1-n/2)\Gamma(1+n/2)}{\Gamma(n/2)\Gamma(2)} \left[-q^2 x(1-x)\right]^{-1+n/2}. \tag{3.18}$$

Similarly, the integral over $d^n p$ of the second term in the square brackets in the second line of Eq. (3.15) corresponds to the parameters $a = 0$, $b = 2$, and $M^2 = -q^2 x(1-x)$ in Eq. (3.16), that yields

$$\int \frac{d^n p}{\left[p^2 - q^2 x(1-x)\right]^2} = \pi^{n/2} \frac{\Gamma(2-n/2)}{\Gamma(2)} \left[-q^2 x(1-x)\right]^{-2+n/2}. \tag{3.19}$$

Hence, by virtue of Eqs. (3.18), (3.19) the regularized hadronic vacuum polarization tensor (3.15) can be represented as

$$\mathrm{reg}\Pi^{(0)}_{\mu\nu}(q^2) = 4iN_c \sum_{f=1}^{n_f} Q_f^2 \mu^{4-n} \frac{\pi^{n/2}}{(2\pi)^n} (-q^2)^{-2+n/2} \int\limits_0^1 [x(1-x)]^{-1+n/2} dx$$

$$\times \left[g_{\mu\nu} \frac{q^2}{2} \frac{\Gamma(1-n/2)\Gamma(1+n/2)}{\Gamma(n/2)} + (2q_\mu q_\nu - q^2 g_{\mu\nu})\Gamma(2-n/2)\right]. \tag{3.20}$$

The integration over dx in Eq. (3.20) can be performed by employing the Eulerian integral of the first kind

$$B(m,n) = \int\limits_0^1 x^{m-1}(1-x)^{n-1} dx = \frac{\Gamma(m)\Gamma(n)}{\Gamma(m+n)}, \tag{3.21}$$

namely,

$$\mathrm{reg}\Pi^{(0)}_{\mu\nu}(q^2) = 8i\left(q_\mu q_\nu - g_{\mu\nu}q^2\right)N_c \sum_{f=1}^{n_f} Q_f^2 \frac{\pi^{n/2}}{(2\pi)^n} \left(\frac{\mu^2}{-q^2}\right)^{2-(n/2)} \frac{\Gamma^2(n/2)}{\Gamma(n)}\Gamma\left(2-\frac{n}{2}\right), \tag{3.22}$$

the relation (3.17) being employed.

As mentioned earlier, the momentum integral in Eq. (3.15) is taken over the noninteger dimension $n = 4 - 2\varepsilon$, so that the hadronic vacuum polarization tensor (3.22) takes the following form:

$$\text{reg}\,\Pi_{\mu\nu}^{(0)}(q^2) = \frac{i}{2\pi^2}\left(q_\mu q_\nu - g_{\mu\nu}q^2\right)N_c\sum_{f=1}^{n_f}Q_f^2\left(4\pi\frac{\mu^2}{-q^2}\right)^\varepsilon\frac{\Gamma^2(2-\varepsilon)}{\Gamma(4-2\varepsilon)}\Gamma(\varepsilon).$$

(3.23)

In turn, the divergent part of the expression (3.23) reveals as the pole terms in ε in the limit $\varepsilon \to 0$. Specifically, by virtue of the relations

$$\lim_{\varepsilon\to 0}\Gamma^2(2-\varepsilon) = 1 + 2\varepsilon(\gamma-1) + \mathcal{O}\left(\varepsilon^2\right),$$

(3.24)

$$\lim_{\varepsilon\to 0}\Gamma(4-2\varepsilon) = 6 + 2\varepsilon(6\gamma-11) + \mathcal{O}\left(\varepsilon^2\right),$$

(3.25)

$$\lim_{\varepsilon\to 0}\Gamma(\varepsilon) = \frac{1}{\varepsilon} - \gamma + \mathcal{O}\left(\varepsilon\right),$$

(3.26)

and

$$\lim_{\varepsilon\to 0}y^\varepsilon = \sum_{k=0}^{\infty}\frac{1}{k!}\varepsilon^k\ln^k y = 1 + \varepsilon\ln y + \mathcal{O}\left(\varepsilon^2\right)$$

(3.27)

the hadronic vacuum polarization tensor (3.23) can be represented as

$$\text{reg}\,\Pi_{\mu\nu}^{(0)}(q^2) = \frac{i}{12\pi^2}\left(q_\mu q_\nu - g_{\mu\nu}q^2\right)N_c\sum_{f=1}^{n_f}Q_f^2\left[\frac{1}{\varepsilon} + \ln(4\pi) - \gamma\right.$$
$$\left. - \ln\left(\frac{-q^2}{\mu^2}\right) + \frac{5}{3}\right],$$

(3.28)

with $\gamma \simeq 0.5772$ being the Euler-Mascheroni constant. Recall that the hadronic vacuum polarization function $\Pi(q^2)$ is defined as (1.28)

$$\Pi_{\mu\nu}(q^2) = \frac{i}{12\pi^2}(q_\mu q_\nu - g_{\mu\nu}q^2)\Pi(q^2).$$

(3.29)

Therefore in the leading order of perturbation theory, namely, in the zeroth order in the strong coupling, the function $\Pi(q^2)$ reads

$$\Pi^{(0)}(q^2) = N_c\sum_{f=1}^{n_f}Q_f^2\left[\frac{1}{\varepsilon} + \ln(4\pi) - \gamma - \ln\left(\frac{-q^2}{\mu^2}\right) + \frac{5}{3}\right].$$

(3.30)

As mentioned in Section 1.1 the common prefactor $N_c\sum_{f=1}^{n_f}Q_f^2$ appearing in front of the functions on hand will be omitted in what follows. It is necessary to outline that, in general, the procedure of regularization of the divergent loop integrals inevitably brings into the theory an additional parameter of the dimension of mass, namely, the renormalization scale μ, which explicitly appears on the right-hand side of Eq. (3.30).

The expression for the subtracted hadronic vacuum polarization function $\Delta\Pi(q^2, q_0^2)$ in the leading order of perturbation theory directly follows from Eq. (3.30), namely,

$$\Delta\Pi_{\text{pert}}^{(0)}(q^2, q_0^2) = -\ln\left(\frac{-q^2}{-q_0^2}\right), \quad -q^2 \to \infty, \tag{3.31}$$

see also Eq. (1.45) and its discussion given in Section 1.2. Evidently, the obtained result (3.31), being rather rough, is valid at high energies only. In particular, the function $\Delta\Pi_{\text{pert}}^{(0)}(q^2, q_0^2)$ diverges at $q^2 \to 0_-$ (irrespective of what value one assigns to the subtraction point q_0^2), which contradicts the corresponding dispersion relation (1.41) and makes the perturbative approximation of $\Pi(q^2)$ inapplicable in the infrared domain (see also Sections 4.3 and 5.1). It is quite obvious that the expression (3.31) does not depend on the choice of the subtraction scheme applied to Eq. (3.30). Specifically, the "minimal subtraction scheme" [205], which is also called "MS-scheme," corresponds to the subtraction from Eq. (3.30) of the divergent term "$1/\varepsilon$" only. Widely used "modified minimal subtraction scheme" [161], which is also called "$\overline{\text{MS}}$-scheme," corresponds to the subtraction from Eq. (3.30) of the divergent term "$1/\varepsilon$" and the constant term "$\ln(4\pi) - \gamma$," that, in turn, leads to

$$\Pi_{\text{pert}}^{(0)}(q^2) = \frac{5}{3} - \ln\left(\frac{-q^2}{\mu^2}\right), \quad -q^2 \to \infty. \tag{3.32}$$

It is also worthwhile to mention the so-called "G-scheme" [208, 209], which differs from the $\overline{\text{MS}}$-scheme by the form of the subtraction constant, as well as the "momentum subtraction scheme" (also known as "MOM-scheme"), which corresponds to the subtraction from the expression on hand (3.30) of its value at a point q_0^2 and thereby yields the result identical to Eq. (3.31). In general, at the higher loop levels the results of QCD perturbative calculations become dependent on the employed subtraction scheme. For example, as mentioned in Section 2.1 the coefficients of perturbative expansion of the renormalization group β function (and, hence, the strong running coupling itself) are scheme-independent at the one-loop and two-loop levels only. The scheme dependence of the hadronic vacuum polarization function and the related functions will be discussed later.

Let us proceed now to the higher orders of perturbation theory. To obtain the expression for the hadronic vacuum polarization function $\Pi(q^2)$ (1.28) in the first order in the strong coupling $\alpha_s = g^2/(4\pi)$ (i.e., at the one-loop level) one has to take into account the contributions of the diagrams displayed in Fig. 3.2. The corresponding calculation yields

$$\Pi_{\text{pert}}^{(1)}(q^2) = \frac{5}{3} - \ln\left(\frac{-q^2}{\mu^2}\right) + \frac{\alpha_s^{(1)}(\mu^2)}{\pi}\left[\frac{55}{12} - 4\zeta(3) - \ln\left(\frac{-q^2}{\mu^2}\right)\right], -q^2 \to \infty, \tag{3.33}$$

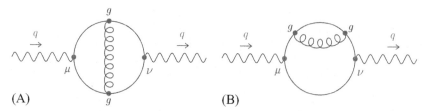

FIG. 3.2 The first order in the strong coupling $\alpha_s = g^2/(4\pi)$ perturbative contributions to the hadronic vacuum polarization function $\Pi(q^2)$ (1.28).

with $\overline{\text{MS}}$-scheme being applied. The expression (3.33) corresponds to the result obtained in early 1970s shortly after the discovery of the QCD asymptotic freedom (see papers [210, 211]). The incorporation of the higher loop strong corrections makes the function $\Pi(q^2)$ rather cumbrous. In particular, at the ℓ-loop level the perturbative approximation of the hadronic vacuum polarization function (1.28) can be represented as ($\ell \geq 1$)

$$\Pi_{\text{pert}}^{(\ell)}(q^2) = \Pi_{\text{pert}}^{(0)}(q^2) + p_{\text{pert}}^{(\ell)}(q^2),$$

$$p_{\text{pert}}^{(\ell)}(q^2) = \sum_{j=1}^{\ell} \left[\frac{\alpha_s^{(\ell)}(\mu^2)}{\pi}\right]^j \sum_{k=0}^{j} \Pi_{j,k} \ln^k\left(\frac{\mu^2}{-q^2}\right). \tag{3.34}$$

In this equation $\Pi_{\text{pert}}^{(0)}(q^2)$ is the leading-order term (3.32), $p_{\text{pert}}^{(\ell)}(q^2)$ denotes the ℓ-loop strong correction, whereas $\Pi_{j,k}$ stand for the corresponding perturbative coefficients. The latter are available up to the fourth order in the strong coupling ($1 \leq \ell \leq 4$) (see papers [212, 213] by Baikov, Chetyrkin, Kuhn, and Rittinger and references therein). Namely, as one can infer from Eq. (3.33) the one-loop ($\ell = 1$) coefficients $\Pi_{1,k}$ read

$$\Pi_{1,0} = \frac{55}{12} - 4\zeta(3), \tag{3.35}$$

$$\Pi_{1,1} = 1. \tag{3.36}$$

In turn, the two-loop ($\ell = 2$) coefficients $\Pi_{2,k}$ take the following form:

$$\Pi_{2,0} = \frac{41927}{864} - \frac{829}{18}\zeta(3) + \frac{25}{3}\zeta(5) + n_f\left[-\frac{3701}{1296} + \frac{19}{9}\zeta(3)\right], \tag{3.37}$$

$$\Pi_{2,1} = \frac{365}{24} - 11\zeta(3) + n_f\left[-\frac{11}{12} + \frac{2}{3}\zeta(3)\right], \tag{3.38}$$

$$\Pi_{2,2} = \frac{1}{8}\left(11 - \frac{2}{3}n_f\right), \tag{3.39}$$

that corresponds to the result known since the late 1970s (see papers [214–216]).

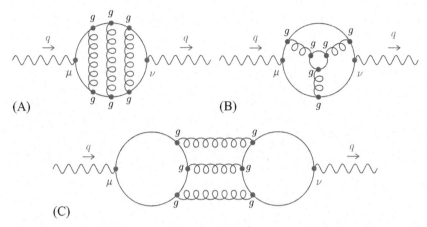

FIG. 3.3 Third order in the strong coupling $\alpha_s = g^2/(4\pi)$ perturbative contributions to the nonsinglet (*plots A and B*) and singlet (*plot C*) parts of the hadronic vacuum polarization function $\Pi(q^2)$ (1.28).

Starting from the three-loop level ($\ell \geq 3$) both nonsinglet and singlet parts of the hadronic vacuum polarization function[1] (labeled, respectively, "NS" and "SI" in what follows) contribute to certain coefficients $\Pi_{j,k}$ (3.34), namely,

$$\Pi_{j,k} = \Pi_{j,k}^{\text{NS}} + \Pi_{j,k}^{\text{SI}} \left(\sum_{f=1}^{n_f} Q_f \right)^2 \left(N_c \sum_{f=1}^{n_f} Q_f^2 \right)^{-1}. \tag{3.40}$$

In this equation Q_f denotes the electric charge of the quark of fth flavor (in units of the elementary charge e) and $N_c = 3$ stands for the number of colors. The singlet part of a coefficient $\Pi_{j,k}$ is assumed to be vanishing unless otherwise explicitly specified. In particular, at the three-loop level ($\ell = 3$)

$$\Pi_{3,0}^{\text{NS}} = \frac{31431599}{41472} - \frac{624799}{864}\zeta(3) + \frac{165}{2}\zeta^2(3) + \frac{55}{48}\zeta(4)$$
$$+ \frac{1745}{96}\zeta(5) - \frac{665}{36}\zeta(7)$$
$$+ n_f \left[-\frac{1863319}{20736} + \frac{174421}{2592}\zeta(3) - \frac{5}{3}\zeta^2(3) - \frac{55}{144}\zeta(4) + \frac{545}{54}\zeta(5) \right]$$
$$+ n_f^2 \left[\frac{196513}{93312} - \frac{809}{648}\zeta(3) - \frac{5}{9}\zeta(5) \right], \tag{3.41}$$

$$\Pi_{3,0}^{\text{SI}} = \frac{2155}{864} - \frac{105}{32}\zeta(3) - \frac{5}{3}\zeta^2(3) - \frac{5}{8}\zeta(4) + \frac{25}{8}\zeta(5), \tag{3.42}$$

---

1. Examples of the corresponding diagrams are displayed in Fig. 3.3.

$$\Pi_{3,1}^{NS} = \frac{87029}{288} - \frac{1103}{4}\zeta(3) + \frac{275}{6}\zeta(5) + n_f\left[-\frac{7847}{216} + \frac{262}{9}\zeta(3) - \frac{25}{9}\zeta(5)\right]$$
$$+ n_f^2\left[\frac{151}{162} - \frac{19}{27}\zeta(3)\right], \tag{3.43}$$

$$\Pi_{3,1}^{SI} = \frac{55}{72} - \frac{5}{3}\zeta(3), \tag{3.44}$$

$$\Pi_{3,2} = \frac{4321}{96} - \frac{121}{4}\zeta(3) + n_f\left[-\frac{785}{144} + \frac{11}{3}\zeta(3)\right] + n_f^2\left[\frac{11}{72} - \frac{1}{9}\zeta(3)\right], \tag{3.45}$$

$$\Pi_{3,3} = \frac{1}{48}\left(11 - \frac{2}{3}n_f\right)^2, \tag{3.46}$$

that corresponds to the result obtained in early 1990s (see papers [217, 218]).

As for the four-loop ($\ell = 4$) coefficients $\Pi_{4,k}$ of the perturbative expansion of the hadronic vacuum polarization function (3.34), the relevant calculation was completed only in early 2010s (see papers [212, 213, 219–221]):

$$\Pi_{4,1}^{NS} = \frac{144939499}{20736} - \frac{5693495}{864}\zeta(3) + \frac{5445}{8}\zeta^2(3) + \frac{65945}{288}\zeta(5) - \frac{7315}{48}\zeta(7)$$
$$+ n_f\left[-\frac{13044007}{10368} + \frac{12205}{12}\zeta(3) - 55\zeta^2(3) + \frac{29675}{432}\zeta(5) + \frac{665}{72}\zeta(7)\right]$$
$$+ n_f^2\left[\frac{1045381}{15552} - \frac{40655}{864}\zeta(3) + \frac{5}{6}\zeta^2(3) - \frac{260}{27}\zeta(5)\right]$$
$$+ n_f^3\left[-\frac{6131}{5832} + \frac{203}{324}\zeta(3) + \frac{5}{18}\zeta(5)\right], \tag{3.47}$$

$$\Pi_{4,1}^{SI} = \frac{5795}{192} - \frac{8245}{144}\zeta(3) - \frac{55}{4}\zeta^2(3) + \frac{2825}{72}\zeta(5)$$
$$+ n_f\left[-\frac{745}{432} + \frac{65}{24}\zeta(3) + \frac{5}{6}\zeta^2(3) - \frac{25}{12}\zeta(5)\right], \tag{3.48}$$

$$\Pi_{4,2}^{NS} = \frac{520175}{384} - \frac{38643}{32}\zeta(3) + \frac{3025}{16}\zeta(5)$$
$$+ n_f\left[-\frac{188521}{768} + \frac{9695}{48}\zeta(3) - \frac{275}{12}\zeta(5)\right]$$
$$+ n_f^2\left[\frac{94693}{6912} - \frac{257}{24}\zeta(3) + \frac{25}{36}\zeta(5)\right] + n_f^3\left[\frac{151}{648} + \frac{19}{108}\zeta(3)\right], \tag{3.49}$$

$$\Pi_{4,2}^{SI} = \frac{605}{192} - \frac{55}{8}\zeta(3) + n_f\left[-\frac{55}{288} + \frac{5}{12}\zeta(3)\right], \tag{3.50}$$

TABLE 3.1 Numerical Values of the Coefficients $\Pi_{j,1}$ of Perturbative Expansion of the Hadronic Vacuum Polarization Function (3.34) up to the Four-Loop Level ($1 \leq j \leq 4$) (See Eqs. 3.36, 3.38, 3.40, 3.43, 3.44, 3.47, 3.48)

n_f	$\Pi_{1,1}$	$\Pi_{2,1}$	$\Pi_{3,1}$	$\Pi_{4,1}$
0	1.0000	1.9857	18.2427	135.7916
1	1.0000	1.8704	13.6999	97.4659
2	1.0000	1.7551	10.0732	73.2196
3	1.0000	1.6398	6.3710	49.0757
4	1.0000	1.5245	2.5933	25.3185
5	1.0000	1.4092	−0.7189	8.7571
6	1.0000	1.2939	−4.1968	−8.3962

$$\Pi_{4,3} = -\frac{1331}{16}\zeta(3) + \frac{49775}{384} + n_f\left[-\frac{2263}{96} + \frac{121}{8}\zeta(3)\right]$$
$$+ n_f^2\left[\frac{593}{432} - \frac{11}{12}\zeta(3)\right] + n_f^3\left[-\frac{11}{432} + \frac{1}{54}\zeta(3)\right], \qquad (3.51)$$

$$\Pi_{4,4} = \frac{1}{256}\left(11 - \frac{2}{3}n_f\right)^3. \qquad (3.52)$$

The coefficient $\Pi_{4,0}$ and the higher-order contributions to $\Pi(q^2)$ have not been evaluated yet. The details of the perturbative calculation of the hadronic vacuum polarization function can be found in extensive reviews [207, 212, 213, 222, 223] and references therein.

In Eqs. (3.35)–(3.52) n_f stands for the number of active flavors (see Section 2.1), $\zeta(x)$ denotes the Riemann ζ function (2.8), $\zeta(3) \simeq 1.20206$, $\zeta(4) = \pi^4/90 \simeq 1.08232$, $\zeta(5) \simeq 1.03693$, $\zeta(7) \simeq 1.00835$, and the SU(3) gauge group is assumed. The calculation of coefficients (3.35)–(3.52) was performed for the case of massless quarks, the $\overline{\text{MS}}$ renormalization scheme being applied. As it will be discussed in the next section, it appears that for the commonly commended choice of the renormalization scale μ^2 the perturbative expression for the Adler function (3.57) retains only the terms proportional to the coefficients $\Pi_{j,1}$. The numerical values of the latter are presented in Table 3.1. As one can infer from this table, for the admissible values of the number of active flavors n_f at the higher loop levels the coefficients $\Pi_{j,1}$ (3.34) may acquire rather large values. It is also worthwhile to mention that the coefficients $\Pi_{j,1}$ ($1 \leq j \leq 4$) assume positive values except for the three-loop coefficient $\Pi_{3,1}$ (which acquires negative values for $n_f = 5$ and $n_f = 6$) and

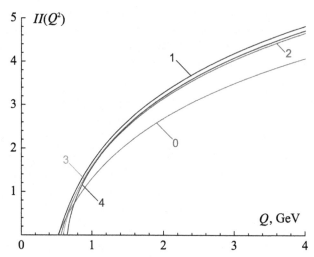

FIG. 3.4 The perturbative approximation of the subtracted hadronic vacuum polarization function $\Delta\Pi_{\text{pert}}^{(\ell)}(-Q_0^2, -Q^2)$ (*solid curves*) at various loop levels ($1 \leq \ell \leq 4$). The *numerical labels* specify the loop level, the parton model prediction (3.31) is marked by "0." The values of parameters: $n_f = 3$ active flavors, $\Lambda = 350\,\text{MeV}$.

the four-loop coefficient $\Pi_{4,1}$ (which acquires negative value for $n_f = 6$; see Table 3.1).

As discussed earlier, the perturbative approximation of the hadronic vacuum polarization function (3.34) diverges at low energies, which makes it incompatible with the dispersion relation (1.41) and bounds the applicability range of $\Pi_{\text{pert}}^{(\ell)}(q^2)$ to the high energies. This issue is illustrated in Fig. 3.4, which displays the perturbative approximation of the subtracted hadronic vacuum polarization function $\Delta\Pi(q^2, q_0^2)$ at various loop levels, indicated by the numerical labels, including the parton model prediction (3.31) labeled by "0." It has to be noted here that because of the infrared divergence of the function $\Pi_{\text{pert}}^{(\ell)}(q^2)$ the latter cannot be subtracted at the point $q_0^2 = 0$ (see also Sections 4.3 and 5.1). Basically, neither the calculation of the higher loop perturbative contributions nor the incorporation of the corrections due to the quark masses[2] enables one to obviate the aforementioned difficulties and obtain a physically sound expression for $\Pi(q^2)$ applicable in the entire energy range. Even if one goes beyond the purely perturbative contributions to the hadronic vacuum polarization function and additionally accounts for the terms corresponding to the diagrams of the type displayed in Fig. 3.5, this does not make the case qualitatively different from that described earlier. Specifically, such diagrams, being generated by the operator product expansion of $\Pi_{\mu\nu}(q^2)$ (1.28), are

---

2. For a discussion of this topic see papers [212, 224–228] and references therein.

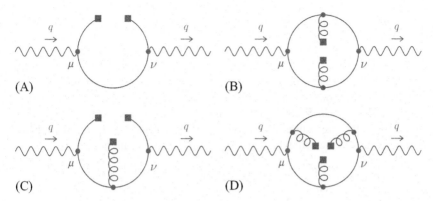

FIG. 3.5 Vacuum condensate contributions to the hadronic vacuum polarization function: quark condensate (*plot A*), gluon condensate (*plot B*), quark-gluon mixed condensate (*plot C*), and three-gluon condensate (*plot D*).

associated with the so-called "vacuum condensate" contributions to $\Pi(q^2)$. The latter constitute the nonlogarithmic power-type corrections of the form of $K_n/(q^2)^n$ with K_n being the phenomenological parameters (constants) of the proper dimensions. This method was proposed in late 1970s by Shifman, Vainshtein, and Zakharov in the works [229–231] and its detailed discussion can be found in the papers [232–253] and references therein. It is also worthwhile to mention that over past two decades there has been considerable progress in the study of the hadronic vacuum polarization function on the lattice that provides a valuable insight into its behavior at low energies (see, in particular, papers [254–277]).

3.2 ADLER FUNCTION WITHIN PERTURBATIVE APPROACH

As it was delineated in Section 1.3, the Adler function $D(Q^2)$ (1.46) is defined as the logarithmic derivative of the hadronic vacuum polarization function $\Pi(q^2)$ (1.28), with $Q^2 = -q^2 \geq 0$ being the spacelike kinematic variable (see paper [100]). The perturbative approximation of the Adler function $D_{\text{pert}}^{(\ell)}(Q^2)$ can be directly derived from the corresponding expression $\Pi_{\text{pert}}^{(\ell)}(q^2)$ (3.34) studied in the previous section. In particular, in the leading order of perturbation theory (i.e., in the zeroth order in the strong coupling) Eqs. (1.46), (3.32) imply that (as mentioned earlier, the common prefactor $N_c \sum_{f=1}^{n_f} Q_f^2$ is omitted throughout)

$$D_{\text{pert}}^{(0)}(Q^2) = -\frac{d}{d\ln Q^2}\Pi_{\text{pert}}^{(0)}(-Q^2) = \frac{d}{d\ln Q^2}\ln\left(\frac{Q^2}{\mu^2}\right) = 1, \quad Q^2 \to \infty,$$

$$(3.53)$$

see also Eq. (1.53) and its discussion given in Section 1.3. It is quite obvious that the expression (3.53), despite the fact that it contains no unphysical singularities that makes it formally applicable in the entire energy range, is rather rough and is valid in the ultraviolet domain only.

Let us address now the Adler function (1.46) in the higher orders of perturbation theory. In particular, to calculate the function $D(Q^2)$ in the first order in the strong coupling $\alpha_s = g^2/(4\pi)$ (or at the one-loop level) one has to take into account the contributions to the hadronic vacuum polarization function $\Pi(q^2)$, which correspond to the diagrams displayed in Fig. 3.2. As specified in the previous section, eventually this leads to Eq. (3.33), which, in turn, results in the following expression for the Adler function:

$$D_{\text{pert}}^{(1)}(Q^2) = -\frac{d}{d\ln Q^2}\Pi_{\text{pert}}^{(1)}(-Q^2) = \left[1 + \frac{1}{\pi}\alpha_s^{(1)}(\mu^2)\right]\frac{d}{d\ln Q^2}\ln\left(\frac{Q^2}{\mu^2}\right)$$

$$= 1 + \frac{1}{\pi}\alpha_s^{(1)}(Q^2), \quad Q^2 \to \infty. \tag{3.54}$$

In the second line of this equation the renormalization scale μ^2 is, as usual, set to be equal to Q^2, $\alpha_s^{(1)}(Q^2) = 4\pi/[\beta_0 \ln(Q^2/\Lambda^2)]$ is the one-loop perturbative strong running coupling (2.19), $\beta_0 = 11 - 2n_f/3$ stands for the one-loop coefficient of perturbative expansion of the renormalization group β function (2.3), and Λ denotes the QCD scale parameter (see Chapter 2 for the details). As one can infer from Eq. (3.54), the incorporation of the strong corrections brings the unphysical singularities into the expression for the Adler function $D(Q^2)$, that makes it inapplicable at low energies.

At the higher loop levels (i.e., for $\ell \geq 2$) the perturbative approximation of the Adler function is determined by Eqs. (1.46), (3.34). At the ℓ-loop level the latter yields

$$D_{\text{pert}}^{(\ell)}(Q^2) = D_{\text{pert}}^{(0)}(Q^2) - \frac{d}{d\ln Q^2}\sum_{j=1}^{\ell}\left[\frac{\alpha_s^{(\ell)}(\mu^2)}{\pi}\right]^j\sum_{k=0}^{j}\Pi_{j,k}\ln^k\left(\frac{\mu^2}{Q^2}\right)$$

$$= D_{\text{pert}}^{(0)}(Q^2) + \sum_{j=1}^{\ell}\left[\frac{\alpha_s^{(\ell)}(\mu^2)}{\pi}\right]^j\sum_{k=1}^{j}(-1)^{k-1}k\Pi_{j,k}\ln^{k-1}\left(\frac{Q^2}{\mu^2}\right).$$

$$\tag{3.55}$$

Then the native choice of the renormalization scale $\mu^2 = Q^2$ eliminates[3] all the terms proportional to the nonvanishing powers of $\ln(Q^2/\mu^2)$ in the second line of Eq. (3.55) and makes the strong coupling α_s running, that is, dependent on the spacelike kinematic variable Q^2, namely,

$$D_{\text{pert}}^{(\ell)}(Q^2) = D_{\text{pert}}^{(0)}(Q^2) + \sum_{j=1}^{\ell}\frac{\Pi_{j,1}}{\pi^j}\left[\alpha_s^{(\ell)}(Q^2)\right]^j, \quad Q^2 \to \infty. \tag{3.56}$$

---

3. In other words, the native choice of the renormalization scale $\mu^2 = Q^2$ retains only the terms proportional to the coefficients $\Pi_{j,1}$ in the second line of Eq. (3.55).

In this equation $D_{\text{pert}}^{(0)}(Q^2) = 1$ is the leading-order term (3.53), the explicit expressions for the available coefficients $\Pi_{j,1}$ ($1 \le j \le 4$) are given in the previous section (their numerical values are listed in Table 3.1), and $\alpha_s^{(\ell)}(Q^2)$ is the ℓ-loop strong running coupling (see Chapter 2 and Appendix A for the details). It is worthwhile to mention here that for another choice of the renormalization scale[4] (e.g., $\mu^2 = cQ^2$ with $c \ne 1$ being a positive constant) all the terms appearing in the second line of Eq. (3.55) contribute to the perturbative approximation of the Adler function.

The expression for the Adler function $D_{\text{pert}}^{(\ell)}(Q^2)$ (3.56) constitutes the perturbative power series in the strong running coupling $\alpha_s^{(\ell)}(Q^2)$. It should be emphasized that the approximation of $D(Q^2)$ in the form of Eq. (3.56) is only valid in the ultraviolet asymptotic, whereas in the infrared domain $D_{\text{pert}}^{(\ell)}(Q^2)$ contains unphysical singularities in the kinematic variable Q^2. Basically, the divergence of the perturbative approximation of the Adler function (3.56) at low energies contradicts the general principles of the theory, is incompatible with the dispersion relation (1.47), and makes $D_{\text{pert}}^{(\ell)}(Q^2)$ inapplicable to the study of the relevant strong interaction processes in the infrared domain that constitutes an intrinsic obstacle of the perturbative approach to QCD.

As discussed in Chapter 2 for practical purposes it proves to be convenient to deal with the so-called perturbative QCD couplant $a_s^{(\ell)}(Q^2)$, which differs from the strong running coupling $\alpha_s^{(\ell)}(Q^2)$ by a numerical factor, which depends on the number of active flavors n_f, specifically, $a_s^{(\ell)}(Q^2) = \alpha_s^{(\ell)}(Q^2)\beta_0/(4\pi)$, where $\beta_0 = 11 - 2n_f/3$ is the first coefficient of the β function perturbative expansion (2.3). It is worth noting that at the one-loop level ($\ell = 1$) the QCD couplant acquires a particularly simple form, namely, $a_s^{(1)}(Q^2) = 1/\ln(Q^2/\Lambda^2)$ with Λ being the QCD scale parameter. The explicit expressions for the QCD couplant at the higher loop levels are given in Chapter 2 and Appendix A. Thus at the ℓ-loop level ($\ell \ge 1$) the perturbative approximation of the Adler function (3.56) can be represented in the following form:

$$D_{\text{pert}}^{(\ell)}(Q^2) = D_{\text{pert}}^{(0)}(Q^2) + d_{\text{pert}}^{(\ell)}(Q^2), \quad d_{\text{pert}}^{(\ell)}(Q^2) = \sum_{j=1}^{\ell} d_j \left[a_s^{(\ell)}(Q^2) \right]^j,$$

$$d_j = \Pi_{j,1} \left(\frac{4}{\beta_0} \right)^j, \tag{3.57}$$

with $D_{\text{pert}}^{(0)}(Q^2) = 1$ being the leading-order term (3.53). As mentioned earlier the coefficients d_j of the perturbative expansion (3.57) are known up to the fourth order in the strong coupling ($1 \le j \le 4$). The one-loop coefficient $d_1 = 4/\beta_0$ is

---

4. For a detailed discussion of the issue of the renormalization scale setting in perturbative QCD calculations, see recent reviews [278, 279] and references therein.

TABLE 3.2 Numerical Values of the Coefficients $d_j = \Pi_{j,1}(4/\beta_0)^j$ of the Adler Function Perturbative Expansion (3.57) up to the Four-Loop Level ($1 \le j \le 4$) (See Eqs. 3.36, 3.38, 3.40, 3.43, 3.44, 3.47, 3.48)

n_f	d_1	d_2	d_3	d_4
0	0.3636	0.2626	0.8772	2.3743
1	0.3871	0.2803	0.7946	2.1884
2	0.4138	0.3005	0.7137	2.1466
3	0.4444	0.3239	0.5593	1.9149
4	0.4800	0.3513	0.2868	1.3440
5	0.5217	0.3836	−0.1021	0.6489
6	0.5714	0.4225	−0.7831	−0.8952

scheme-independent, whereas the higher-order coefficients d_j ($j \ge 2$) depend on the employed renormalization scheme. In what follows all the scheme-dependent perturbative coefficients will be assumed to correspond to the $\overline{\text{MS}}$-scheme unless otherwise specified.

The numerical values of the currently available coefficients d_j (3.57) are presented in Table 3.2. As one can infer from this table for the admissible values of the number of active flavors n_f the coefficients d_j (contrary to the coefficients $\Pi_{j,1}$) do not acquire large values. Similar to the coefficients $\Pi_{j,1}$ (3.34) the coefficients d_1 and d_2 assume positive values for $0 \le n_f \le 6$ and so do the coefficients d_3 (except for $n_f = 5$ and $n_f = 6$) and d_4 (except for $n_f = 6$). As discussed in the previous section, starting from the three-loop level the diagrams of both nonsinglet (labeled "NS") and singlet (labeled "SI") types (see Fig. 3.3) contribute to the hadronic vacuum polarization function $\Pi(q^2)$ (1.28). Hence, likewise the coefficients $\Pi_{j,k}$ (3.40) the coefficients d_j (3.57) are split into two parts for $j \ge 3$, specifically,

$$d_j = d_j^{\text{NS}} + d_j^{\text{SI}} \left(\sum_{f=1}^{n_f} Q_f \right)^2 \left(N_c \sum_{f=1}^{n_f} Q_f^2 \right)^{-1}, \quad d_j^{\text{NS}} = \Pi_{j,1}^{\text{NS}} \left(\frac{4}{\beta_0} \right)^j,$$

$$d_j^{\text{SI}} = \Pi_{j,1}^{\text{SI}} \left(\frac{4}{\beta_0} \right)^j, \tag{3.58}$$

where $N_c = 3$ denotes the number of colors and Q_f stands for the electric charge of the quark of fth flavor (in units of the elementary charge e). As one can infer from Table 3.3, at the currently accessible loop levels ($1 \le \ell \le 4$) and for the admissible values of the number of active flavors n_f the nonsinglet part of the coefficient d_j dominates over its singlet part, except for the case of $\ell = 4$ and $n_f = 6$. Note that though the coefficients $\Pi_{j,1}$ of the Adler

TABLE 3.3 The Relative Weight of the Singlet Part $(1 + |1 - d_j/d_j^{NS}|^{-1})^{-1} \times$ 100% in the Coefficients d_j of the Adler Function Perturbative Expansion (3.57) at First Four Loop Levels $(1 \leq j \leq 4)$

n_f	$j = 1$	$j = 2$	$j = 3$	$j = 4$
0	0.00%	0.00%	0.00%	0.00%
1	0.00%	0.00%	2.84%	5.28%
2	0.00%	0.00%	0.81%	1.47%
3	0.00%	0.00%	0.00%	0.00%
4	0.00%	0.00%	5.65%	7.03%
5	0.00%	0.00%	5.22%	4.69%
6	0.00%	0.00%	5.91%	34.25%

function perturbative expansion (3.56) have not been calculated beyond the four-loop level yet, there is a number of attempts to estimate[5] their numerical values for $j > 4$ in the framework of various optimization schemes. Among the latter it is certainly worthwhile to mention the "method of effective charges" developed by Grunberg in papers [283–286], the "principle of minimum sensitivity" suggested by Stevenson in papers [287–295], the "scheme-invariant perturbation theory" proposed by Dhar and Gupta in papers [296, 297], the "Brodsky-Lepage-Mackenzie method" developed by the named authors in their paper [298] (see also various modifications of this method suggested in papers [299–303]), and a recently proposed by Brodsky, Di Giustino, and Wu "principle of maximum conformality" [304–311] (see also extensive reviews [278, 279] and references therein).

Thus the perturbative expression for the Adler function $D_{\text{pert}}^{(\ell)}(Q^2)$ (3.57) constitutes the sum of the leading-order term $D_{\text{pert}}^{(0)}(Q^2)$ (3.53), which formally contains no unphysical singularities in Q^2, but represents a rather rough approximation, and the corresponding strong correction $d_{\text{pert}}^{(\ell)}(Q^2)$, which diverges at low energies due to the unphysical singularities of the strong running coupling. As mentioned earlier the presence of such singularities is incompatible with the pertinent dispersion relation (1.47) and makes the perturbative expression for the Adler function (3.57) inapplicable in the infrared domain. This issue is illustrated in Fig. 3.6, which displays the perturbative approximation of the Adler function (3.57) at various loop levels, which are indicated by the corresponding numerical labels (the parton model prediction (3.53) is labeled by "0"). Similar to the case of the hadronic vacuum polarization function

5. See, for example, papers [108, 280–282], as well as Section 6.3 and references therein.

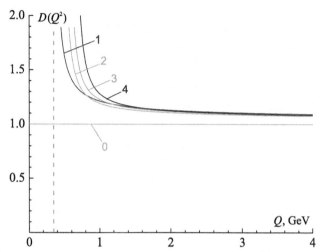

FIG. 3.6 The perturbative approximation of the Adler function $D_{\text{pert}}^{(\ell)}(Q^2)$ (Eq. 3.57, *solid curves*) at various loop levels ($1 \le \ell \le 4$). *Horizontal solid line* represents the parton model prediction $D_{\text{pert}}^{(0)}(Q^2)$ (3.53), whereas the *vertical dashed line* marks the location of the unphysical "ghost pole" ($Q^2 = \Lambda^2$). The *numerical labels* specify the loop level. The values of parameters: $n_f = 3$ active flavors, $\Lambda = 350\,\text{MeV}$.

discussed in the previous section, neither the calculation of the higher-order perturbative contributions nor the incorporation of the corresponding nonlogarithmic "vacuum condensate" corrections (which, nonetheless, supply certain additional input) enable one to circumvent the intrinsic difficulties of the perturbative approach and to obtain the expression for $D(Q^2)$ applicable in the entire energy range. It is also worthwhile to mention that over the past decades the Adler function was extensively studied in the framework of various approaches (see, in particular, papers [92, 140, 141, 312–329]). Additionally, a valuable part of the information on the behavior of the function $D(Q^2)$ in the infrared domain is provided by the lattice studies performed, for example, in the papers [266, 271–275]. The detailed description of the aforementioned methods can be found in the quoted references.

3.3 *R*-RATIO WITHIN PERTURBATIVE APPROACH

It has to be outlined from the very beginning that the theoretical description of the *R*-ratio of electron-positron annihilation into hadrons essentially differs from that of the hadronic vacuum polarization function $\Pi(q^2)$ and the Adler function $D(Q^2)$ studied, respectively, in Sections 3.1 and 3.2. Basically, this is caused by the fact that both $\Pi(q^2)$ and $D(Q^2)$ are the functions of the spacelike kinematic variable $-q^2 = Q^2 \ge 0$, and thereby their behavior at high energies $-q^2 = Q^2 \to \infty$ is accessible within QCD perturbation theory. On the contrary,

$R(s)$ is a function of the timelike kinematic variable $s = q^2 \geq 0$ that makes the perturbative approach directly inapplicable[6] to the study of the latter. As noted in Section 1.1, it is the relevant dispersion relation that enables one to describe the R-ratio of electron-positron annihilation into hadrons in a self-consistent way. In this section only the most salient features of the behavior of the function $R(s)$ at high energies will be highlighted, whereas the detailed study of this matter will be presented in Chapters 4–6 and Appendix C.

Let us begin by addressing the R-ratio of electron-positron annihilation into hadrons in the leading order of perturbation theory, namely, in the zeroth order in the strong coupling. In this case the relation (1.40) enables one to obtain the parton model prediction $R_{\text{pert}}^{(0)}(s)$ from the corresponding expression for the hadronic vacuum polarization function $\Pi_{\text{pert}}^{(0)}(q^2)$ (3.32), specifically,

$$
\begin{aligned}
R_{\text{pert}}^{(0)}(s) &= \frac{1}{2\pi i} \lim_{\varepsilon \to 0_+} \left[-\ln\left(\frac{-s - i\varepsilon}{\mu^2}\right) + \ln\left(\frac{-s + i\varepsilon}{\mu^2}\right) \right] \\
&= \frac{1}{2\pi i} \left[-\ln\left(\frac{s}{\mu^2}\right) + i\pi + \ln\left(\frac{s}{\mu^2}\right) + i\pi \right] = 1, \quad s \to \infty, \quad (3.59)
\end{aligned}
$$

see also Eq. (1.32) and its discussion given in Section 1.1. As mentioned earlier, the common prefactor $N_c \sum_{f=1}^{n_f} Q_f^2$ is omitted throughout.

Equivalently, the function $R(s)$ can also be derived from the expression for the Adler function $D(Q^2)$ by making use of the relation (1.50). The possible integration contours in Eq. (1.50) corresponding to the massless limit are displayed in Fig. 3.7. In this case the parton model prediction $D_{\text{pert}}^{(0)}(Q^2)$ (3.53) and the relation (1.50) imply that

$$
\begin{aligned}
R_{\text{pert}}^{(0)}(s) &= \frac{1}{2\pi i} \lim_{\varepsilon \to 0_+} \int_{s+i\varepsilon}^{s-i\varepsilon} D_{\text{pert}}^{(0)}(-\zeta) \frac{d\zeta}{\zeta} \\
&= \frac{1}{2\pi i} \lim_{\varepsilon \to 0_+} \left[\int_{s+i\varepsilon}^{\infty+i\varepsilon} \frac{d\zeta}{\zeta} + \int_{C_\infty} \frac{d\zeta}{\zeta} + \int_{\infty-i\varepsilon}^{s-i\varepsilon} \frac{d\zeta}{\zeta} \right], \quad (3.60)
\end{aligned}
$$

with the integration contour shown in Fig. 3.7A being employed. Since the first and third terms in the square brackets in the second line of Eq. (3.60) cancel each other, after the change of the integration variable $\zeta = r\exp(i\varphi)$ in the second term one arrives at

$$
R_{\text{pert}}^{(0)}(s) = \frac{1}{2\pi i} \lim_{\varepsilon \to 0_+} \lim_{r \to \infty} \int_{\varepsilon}^{2\pi - \varepsilon} \frac{r i e^{i\varphi}}{r e^{i\varphi}} d\varphi = 1, \quad (3.61)
$$

---

6. For a discussion of this issue see, in particular, papers [105–108, 135–138].

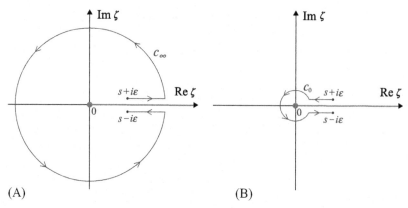

FIG. 3.7 Possible integration contours in the complex ζ-plane in Eq. (1.50) in the massless limit. The cut $\zeta \geq 0$ of the Adler function $D(-\zeta)$ is shown along the positive semiaxis of real ζ.

that coincides with Eq. (3.59). It is straightforward to verify that the use of the integration contour displayed in Fig. 3.7B in Eq. (1.50) leads to a result identical to Eq. (3.61). Evidently, the parton model prediction $R_{\text{pert}}^{(0)}(s)$ (3.61), though it contains no unphysical singularities, represents a rather rough approximation of the R-ratio, which is valid at high energies only.

As for the higher loop levels the continuation (1.50) of the perturbative approximation of the Adler function $D_{\text{pert}}^{(\ell)}(Q^2)$ (3.57) into the timelike domain leads to an essential distortion of its form even in the ultraviolet asymptotic. In particular, in the massless limit at the one-loop level such continuation results in

$$R^{(1)}(s) = 1 + d_1 \left\{ \frac{1}{2} - \frac{1}{\pi} \arctan\left[\frac{\ln(s/\Lambda^2)}{\pi}\right] \right\}, \qquad (3.62)$$

that certainly differs from Eq. (3.54) (see also papers [105, 330, 331] as well as Eqs. 6.2, 6.3). In Eq. (3.62) $d_1 = 4/\beta_0$ denotes the one-loop perturbative coefficient of the Adler function (see Section 3.2), $\beta_0 = 11 - 2n_f/3$ stands for the one-loop perturbative coefficient of the β function (see Section 2.1), and n_f is the number of active flavors. At the same time the right-hand side of Eq. (3.62), being reexpanded for $\ln(s/\Lambda^2) > \pi$, can be represented as the sum of the "naive" form of the function $R(s)$ and the corrections, which appear to be beyond the accuracy of the one-loop level, namely,

$$R^{(1)}(s) \simeq R_{\text{naive}}^{(1)}(s) + \mathcal{O}\left(\frac{1}{\ln^3 w}\right), \quad w = \frac{s}{\Lambda^2}, \quad \sqrt{s} > \Lambda \exp\left(\frac{\pi}{2}\right) \simeq 4.81\Lambda,$$
$$(3.63)$$

where

$$R_{\text{naive}}^{(\ell)}(s) = D_{\text{pert}}^{(\ell)}(|s|) = 1 + \sum_{j=1}^{\ell} d_j \left[a_s^{(\ell)}(|s|)\right]^j, \quad s = q^2 = -Q^2 \geq 0, \quad (3.64)$$

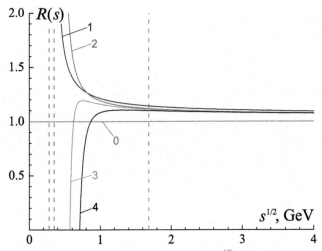

FIG. 3.8 The perturbative approximation of the function $R_{\text{pert}}^{(\ell)}(s)$ (Eq. 3.67, *solid curves*) at various loop levels ($1 \leq \ell \leq 4$). *Horizontal solid line* represents the parton model prediction (3.61), whereas the *vertical dashed lines* mark the locations of (from left to right) the kinematic threshold at $\sqrt{s} = 2m_\pi$, the one-loop perturbative pole at $\sqrt{s} = \Lambda$, and the boundary of the convergence range of the approximation (3.67) at $\sqrt{s} = \exp(\pi/2)\Lambda \simeq 4.81\Lambda$. The *numerical labels* specify the loop level. The values of parameters: $n_f = 3$ active flavors, $\Lambda = 350\,\text{MeV}$.

and the perturbative approximation of the Adler function $D_{\text{pert}}^{(\ell)}(Q^2)$ is specified in Eq. (3.57). However, starting from the three-loop level (i.e., for $\ell \geq 3$) the terms, which constitute the difference between $R_{\text{naive}}^{(\ell)}(s)$ (3.64) and the reexpansion of $R^{(\ell)}(s)$ for $\ln(s/\Lambda^2) > \pi$ cannot be neglected anymore, for example

$$R^{(3)}(s) \simeq R_{\text{naive}}^{(3)}(s) - \frac{\pi^2}{3}\frac{d_1}{\ln^3 w} + \mathcal{O}\left(\frac{1}{\ln^4 w}\right), \quad s \to \infty, \qquad (3.65)$$

see also Eq. (6.15). As it will be discussed in Chapters 5 and 6 at any given loop level the reexpansion of the function $R^{(\ell)}(s)$ at high energies can be reduced to the form of a perturbative power series in the naive continuation of the strong running coupling $\alpha_s^{(\ell)}(Q^2)$ into the timelike domain, specifically,

$$R^{(\ell)}(s) \simeq R_{\text{pert}}^{(\ell)}(s) + \mathcal{O}\left[\frac{1}{\ln^{\ell+1}(s/\Lambda^2)}\right], \quad \sqrt{s} > \Lambda \exp\left(\frac{\pi}{2}\right) \simeq 4.81\Lambda.$$

$$(3.66)$$

In this equation $R_{\text{pert}}^{(\ell)}(s)$ is the ℓ-loop perturbative approximation of the R-ratio of electron-positron annihilation into hadrons

$$R_{\text{pert}}^{(\ell)}(s) = R_{\text{pert}}^{(0)}(s) + r_{\text{pert}}^{(\ell)}(s), \quad r_{\text{pert}}^{(\ell)}(s) = \sum_{j=1}^{\ell} r_j \left[a_s^{(\ell)}(|s|)\right]^j, \quad r_j = d_j - \delta_j,$$

$$(3.67)$$

where $R_{\text{pert}}^{(0)}(s) = 1$ stands for the leading-order term (3.61), d_j denote the coefficients of the Adler function perturbative expansion (3.57), and the coefficients δ_j embody the contributions due to the so-called π^2-terms (see Chapters 5 and 6, and Appendix C for details). The first two coefficients δ_1 and δ_2 vanish,[7] whereas at the higher loop levels (i.e., for $j \geq 3$) δ_j appear to dominate over d_j, which is illustrated in Tables 3.2, 6.1, 6.2, and 6.4, as well as in Figs. 3.6 and 3.8. In particular, as one can infer from these figures the three-loop and four-loop expressions for the Adler function (3.57) and the R-ratio (3.67) behave in a rather different way when $|q^2|$ approaches infrared domain. It is also worthwhile to note that the perturbative approximation of the R-ratio (3.67) diverges at low energies due to the unphysical singularities of the strong running coupling, which is physically unjustified and contradicts the pertinent experimental observations.

Thus the proper continuation of theoretical results from spacelike to timelike domain leads to an essential distortion of the original perturbative power series. In turn, this plays a substantial role in the study of the R-ratio of electron-positron annihilation into hadrons and the related quantities in the entire energy range. A detailed discussion of this issue will be presented in the next chapters.

---

7. That makes the perturbative approximation of the R-ratio (3.67) identical to its naive form (3.64) at the first two loop levels.

Chapter 4

Dispersive Approach to QCD

Many strong interaction processes are governed by the hadronic vacuum polarization function $\Pi(q^2)$. The QCD perturbation theory provides only a part of the information on this function, specifically, its behavior at high energies. Another part of the information about $\Pi(q^2)$ and the related functions, being based on the kinematic restrictions on pertinent physical processes, is provided by the relevant dispersion relations. The dispersive approach to QCD merges these two inputs in a self-consistent way, gets rid of some of the intrinsic obstacles of perturbative approach, extends its range of applicability toward the infrared domain, and properly accounts for the effects due to continuation of spacelike theoretical results into the timelike domain.

4.1 UNIFIED INTEGRAL REPRESENTATIONS FOR FUNCTIONS $\Pi(q^2)$, $R(s)$, AND $D(Q^2)$

Let us begin by briefly recalling the basic dispersion relations for the hadronic vacuum polarization function $\Pi(q^2)$, the function $R(s)$, and the Adler function $D(Q^2)$. As discussed in Chapter 1 the dispersion relation for the hadronic vacuum polarization function (1.41) can be derived by making use of only its definition (1.28), its ultraviolet behavior (1.45), and the respective kinematic restrictions:

$$\Delta\Pi(q^2, q_0^2) = (q^2 - q_0^2) \int_{m^2}^{\infty} \frac{R(s)}{(s - q^2)(s - q_0^2)} \, ds, \qquad (4.1)$$

where $\Delta\Pi(q^2, q_0^2) = \Pi(q^2) - \Pi(q_0^2)$ and q_0^2 stands for the subtraction point. In Eq. (4.1) the points q^2 and q_0^2 can be located anywhere in the complex q^2-plane except for the physical cut $q^2 \geq m^2$ along the positive semiaxis of real q^2. In turn, the Adler function [100], being defined as the logarithmic derivative of the hadronic vacuum polarization function (1.46)

$$D(Q^2) = -\frac{d\Pi(-Q^2)}{d \ln Q^2} \qquad (4.2)$$

Strong Interactions in Spacelike and Timelike Domains. http://dx.doi.org/10.1016/B978-0-12-803439-2.00004-1

is independent of the aforementioned subtraction point q_0^2. The corresponding dispersion relation (1.47) immediately follows from Eqs. (4.1), (4.2), namely [100],

$$D(Q^2) = Q^2 \int_{m^2}^{\infty} \frac{R(s)}{(s + Q^2)^2} \, ds. \tag{4.3}$$

The function $R(s)$ entering Eqs. (4.1), (4.3) is determined as the discontinuity of the hadronic vacuum polarization function $\Pi(q^2)$ across the physical cut (1.40)

$$R(s) = \frac{1}{2\pi i} \lim_{\varepsilon \to 0_+} \Delta \Pi(s + i\varepsilon, s - i\varepsilon) \tag{4.4}$$

and is identified with the R-ratio of electron-positron annihilation into hadrons (1.1). Then the integration of definition (4.2) in finite limits results in the inverse relation (1.51) between functions $\Pi(q^2)$ and $D(Q^2)$, specifically [139],

$$\Delta \Pi(-Q^2, -Q_0^2) = - \int_{Q_0^2}^{Q^2} D(\zeta) \frac{d\zeta}{\zeta}. \tag{4.5}$$

And finally the inverse relation (1.50) between functions $R(s)$ and $D(Q^2)$ directly follows from Eqs. (4.4), (4.5), namely [105, 106]

$$R(s) = \frac{1}{2\pi i} \lim_{\varepsilon \to 0_+} \int_{s+i\varepsilon}^{s-i\varepsilon} D(-\zeta) \frac{d\zeta}{\zeta}. \tag{4.6}$$

In Eqs. (4.1)–(4.6) $Q^2 = -q^2 \geq 0$ and $s = q^2 \geq 0$ are the spacelike and timelike kinematic variables, respectively, whereas $m = 2m_\pi$ denotes the total mass of the respective lightest allowed hadronic final state. In Eq. (4.6) the integration contour in the complex ζ-plane lies in the region of analyticity of the integrand. It is also necessary to outline that, as discussed in Chapters 1 and 3, the theoretical analysis of the strong interaction processes in the spacelike domain (e.g., the study of the behavior of functions $\Pi(q^2)$ and $D(Q^2)$ at high energies $Q^2 = -q^2 \to \infty$) can be performed in the framework of QCD perturbation theory. However, the latter is not directly applicable to the analysis of hadron dynamics in the timelike domain. In particular, the self-consistent description of the R-ratio of electron-positron annihilation into hadrons is inherently based on the dispersion relations (4.1)–(4.6) and cannot be performed without the latter.

Eqs. (4.1)–(4.6) constitute the complete set of relations, which express the functions $\Pi(q^2)$, $R(s)$, and $D(Q^2)$ in terms of each other. These relations impose stringent physical intrinsically nonperturbative constraints on the functions on hand, which should certainly be accounted for when one comes out of the limits of applicability of perturbation theory. In particular, the dispersion relation (4.1) signifies that the hadronic vacuum polarization function $\Pi(q^2)$ possesses the only[1] cut $q^2 \geq m^2$ along the positive semiaxis of real q^2 and that the points q^2 and q_0^2 can be located anywhere in the complex q^2-plane except for this cut (that turns out to be substantial for the study of the infrared behavior of $\Pi(q^2)$; see Section 5.1). Eqs. (4.4), (4.6) indicate that the function $R(s)$ assumes nonzero values for real s above the threshold only (i.e., for $s \geq m^2$) and accounts for the effects due to a continuation of spacelike theoretical results into the timelike domain that plays a significant role even in the ultraviolet asymptotic $s \to \infty$ (see Section 5.2 and Chapter 6). The dispersion relation (4.3) implies that the Adler function $D(Q^2)$ possesses the only cut $Q^2 \leq -m^2$ along the negative semiaxis of real Q^2 and that $D(Q^2)$ vanishes in the infrared limit $Q^2 \to 0$ (the latter condition holds for $m \neq 0$ only and appears to be lost in the massless limit). In turn, the first of these constraints indicates that the Adler function contains no unphysical singularities, whereas the second one essentially stabilizes its infrared behavior (see Section 5.3).

For practical purposes it proves to be convenient to deal with the unified integral representations, which express the functions on hand in terms of the common spectral density. In particular, this enables one to merge, in a self-consistent way, the aforementioned nonperturbative constrains with corresponding perturbative input, which constitutes an underlying concept of the dispersive approach[2] to QCD [336–338].

The basic dispersion relations for the functions on hand (4.1)–(4.6) satisfy the additivity property, which allows one to handle the leading-order terms[3] $\Pi^{(0)}(q^2)$, $R^{(0)}(s)$, and $D^{(0)}(Q^2)$ separately from the respective strong corrections $p(q^2)$, $r(s)$, and $d(Q^2)$. Let us first derive the integral representation for the strong correction to the function $R(s)$. To achieve this objective it is worthwhile to employ the relation (4.6), which expresses the function $R(s)$ in terms of the Adler function $D(Q^2)$. The possible integration contours in Eq. (4.6) are shown in Fig. 4.1.

For the energies exceeding the hadronic production threshold ($s \geq m^2$) Eq. (4.6) for the strong correction $r(s)$ reads

$$
r(s) = \frac{1}{2\pi i} \lim_{\varepsilon \to 0_+} \left[\int_{s+i\varepsilon}^{\infty+i\varepsilon} d(-\zeta)\frac{d\zeta}{\zeta} + \int_{c_\infty} d(-\zeta)\frac{d\zeta}{\zeta} + \int_{\infty-i\varepsilon}^{s-i\varepsilon} d(-\zeta)\frac{d\zeta}{\zeta} \right], \quad (4.7)
$$

---

1. That implies that the hadronic vacuum polarization function $\Pi(q^2)$ is free of the unphysical singularities.
2. The preliminary formulation of the dispersive approach to QCD was discussed in papers [332–335].
3. That is, the terms of the zeroth order in the strong running coupling.

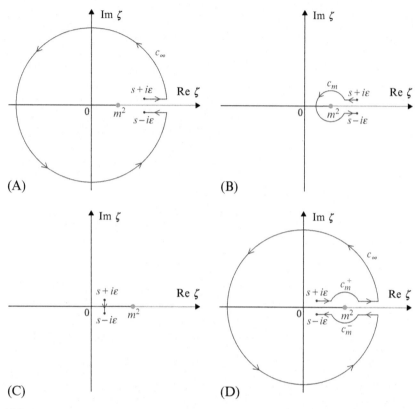

FIG. 4.1 The possible integration contours in the complex ζ-plane in Eq. (4.6) for the energies above $(s \geq m^2$, plots A and B) and below $(0 \leq s < m^2$, plots C and D) the hadronic production threshold. The physical cut $\zeta \geq m^2$ of the Adler function $D(-\zeta)$ (4.2) is shown along the positive semiaxis of real ζ.

the integration contour shown in Fig. 4.1A being employed. It is convenient to change the integration variable $\zeta = \sigma + i\varepsilon$ in the first term in the square brackets on the right-hand side of Eq. (4.7):

$$\int_{s+i\varepsilon}^{\infty+i\varepsilon} d(-\zeta)\frac{d\zeta}{\zeta} = \int_{s}^{\infty} d(-\sigma - i\varepsilon)\frac{d\sigma}{\sigma} + \mathcal{O}(\varepsilon). \tag{4.8}$$

Similarly, the change of the integration variable $\zeta = \sigma - i\varepsilon$ in the third term in the square brackets on the right-hand side of Eq. (4.7) leads to

$$\int_{\infty-i\varepsilon}^{s-i\varepsilon} d(-\zeta)\frac{d\zeta}{\zeta} = -\int_{s}^{\infty} d(-\sigma + i\varepsilon)\frac{d\sigma}{\sigma} + \mathcal{O}(\varepsilon). \tag{4.9}$$

In Eq. (4.6) the integration of the leading-order term of the Adler function $D^{(0)}(Q^2)$ along the circle of infinitely large radius (i.e., along the subcontour c_∞ in Fig. 4.1A) yields a nonvanishing contribution to $R(s)$ (see Section 3.3), whereas the result of integration of the corresponding strong correction $d(Q^2)$ vanishes. In particular, the second term in the square brackets on the right-hand side of Eq. (4.7) gives no contribution due to the QCD asymptotic freedom:

$$\int_{c_\infty} d(-\zeta)\frac{d\zeta}{\zeta} \to \frac{4}{\beta_0}\int_{c_\infty}\frac{1}{\ln(-\zeta/\Lambda^2)}\frac{d\zeta}{\zeta} = \lim_{\varepsilon\to 0_+}\lim_{r\to\infty}\frac{4}{\beta_0}\int_\varepsilon^{2\pi-\varepsilon}\frac{id\varphi}{\ln[-r\exp(i\varphi)]}$$

$$= \lim_{\varepsilon\to 0_+}\lim_{r\to\infty}\frac{4}{\beta_0}\int_{-\pi+\varepsilon}^{\pi-\varepsilon}\frac{id\varphi}{\ln(r)+i\varphi} = 2\pi i\frac{4}{\beta_0}\lim_{r\to\infty}\frac{1}{\pi}\arctan\left(\frac{\pi}{\ln r}\right) = 0.$$

$$(4.10)$$

Thus Eq. (4.7) acquires the form

$$r(s) = \int_s^\infty \rho(\sigma)\frac{d\sigma}{\sigma}, \quad s \ge m^2, \tag{4.11}$$

where

$$\rho(\sigma) = \frac{1}{2\pi i}\lim_{\varepsilon\to 0_+}[d(-\sigma - i\varepsilon) - d(-\sigma + i\varepsilon)] \tag{4.12}$$

is the aforementioned spectral density. As it will be discussed in the next section, it is this quantity that brings the corresponding perturbative input into the unified integral representations for the functions on hand.

It is necessary to outline here that Eq. (4.6) leads to the result independent of the employed integration contour only for the strong correction to the Adler function $d(Q^2)$, which possesses the proper[4] analytic properties in the kinematic variable Q^2. In particular, for the integration contour presented in Fig. 4.1B, Eq. (4.6) acquires the form

$$r(s) = \frac{1}{2\pi i}\lim_{\varepsilon\to 0_+}\left[\int_{s+i\varepsilon}^{m^2+i\varepsilon} d(-\zeta)\frac{d\zeta}{\zeta} + \int_{c_m} d(-\zeta)\frac{d\zeta}{\zeta} + \int_{m^2-i\varepsilon}^{s-i\varepsilon} d(-\zeta)\frac{d\zeta}{\zeta}\right].$$

$$(4.13)$$

Following the very same lines as above, one can cast Eq. (4.13) to

---

4. Namely, for the function $d(Q^2)$, which possesses the only cut $Q^2 \le -m^2$ along the negative semiaxis of real Q^2, as prescribed by the dispersion relation (4.3).

$$r(s) = \frac{1}{2\pi i} \int\limits_{C_m} d(-\zeta)\frac{d\zeta}{\zeta} - \int\limits_{m^2}^{s} \rho(\sigma)\frac{d\sigma}{\sigma}, \tag{4.14}$$

where the spectral density $\rho(\sigma)$ is defined in Eq. (4.12). As it will be shown below, the strong correction to the Adler function $d(Q^2)$, which satisfies Eq. (4.3), can also be represented in the form of Eq. (4.23). In turn, this fact allows one to reduce the first term of Eq. (4.14) to

$$\frac{1}{2\pi i} \int\limits_{C_m} d(-\zeta)\frac{d\zeta}{\zeta} = \frac{1}{2\pi i} \int\limits_{C_m} \frac{d\zeta}{\zeta - m^2} \int\limits_{m^2}^{\infty} \frac{d\sigma}{\sigma}\rho(\sigma)\frac{\sigma - m^2}{\sigma - \zeta} = \int\limits_{m^2}^{\infty} \rho(\sigma)\frac{d\sigma}{\sigma}. \tag{4.15}$$

Hence, Eq. (4.14) takes the form

$$r(s) = \int\limits_{s}^{\infty} \rho(\sigma)\frac{d\sigma}{\sigma}, \quad s \geq m^2, \tag{4.16}$$

which coincides with the result (4.11). The identity of expressions (4.11), (4.16) illustrates the fact that in Eq. (4.6) one can continuously deform the integration contour shown in Fig. 4.1A to the one presented in Fig. 4.1B and vice versa, the validity of representation (4.3) for the strong correction to the Adler function being assumed. It is worthwhile to note here that for the perturbative approximation of the strong correction to the Adler function $d_{\text{pert}}^{(\ell)}(Q^2)$, which possesses infrared unphysical singularities at positive values of real Q^2, such continuous deformation of the integration contour in Eq. (4.6) cannot be performed, see also discussion of this issue in Section 4.2.

At the same time it is straightforward to verify that for the energies below the hadronic production threshold ($0 \leq s < m^2$), the integration of Eq. (4.6) along either of the contours shown in Fig. 4.1C and D yields

$$r(s) = 0, \quad 0 \leq s < m^2. \tag{4.17}$$

Thus Eqs. (4.11), (4.17) imply that the strong correction to the R-ratio of electron-positron annihilation into hadrons acquires the form

$$r(s) = \theta(s - m^2)\int\limits_{s}^{\infty} \rho(\sigma)\frac{d\sigma}{\sigma}, \quad s \geq 0, \tag{4.18}$$

where the spectral density $\rho(\sigma)$ is specified in Eq. (4.12) and $\theta(x)$ stands for the Heaviside unit step function (1.27).

In turn, the strong correction to the hadronic vacuum polarization function $p(q^2)$ can be expressed in terms of the spectral density $\rho(\sigma)$ by making use of Eqs. (4.1), (4.18), namely,

$$\Delta p(q^2, q_0^2) = (q^2 - q_0^2) \int\limits_{m^2}^{\infty} ds \frac{\theta(s - m^2)}{(s - q^2)(s - q_0^2)} \int\limits_{s}^{\infty} \rho(\sigma) \frac{d\sigma}{\sigma}, \tag{4.19}$$

where $\Delta p(q^2, q_0^2) = p(q^2) - p(q_0^2)$. The integration by parts allows one to cast this equation to

$$\Delta p(q^2, q_0^2) = r(s) \ln\left(\frac{s - q^2}{s - q_0^2}\right)\Bigg|_{s=m^2}^{s=\infty} + \int\limits_{m^2}^{\infty} \ln\left(\frac{\sigma - q^2}{\sigma - q_0^2}\right) \rho(\sigma) \frac{d\sigma}{\sigma}. \tag{4.20}$$

By virtue of Eq. (4.18) the first term on the right-hand side of Eq. (4.20) can be represented as

$$r(s) \ln\left(\frac{s - q^2}{s - q_0^2}\right)\Bigg|_{s=m^2}^{s=\infty} = \ln\left(\frac{m^2 - q_0^2}{m^2 - q^2}\right) \int\limits_{m^2}^{\infty} \rho(\sigma) \frac{d\sigma}{\sigma}. \tag{4.21}$$

Therefore the subtracted strong correction to the hadronic vacuum polarization function (4.20) takes the following form

$$\Delta p(q^2, q_0^2) = \int\limits_{m^2}^{\infty} \rho(\sigma) \ln\left(\frac{\sigma - q^2}{\sigma - q_0^2} \frac{m^2 - q_0^2}{m^2 - q^2}\right) \frac{d\sigma}{\sigma}, \tag{4.22}$$

where the spectral density $\rho(\sigma)$ is defined in Eq. (4.12).

And finally the strong correction to the Adler function $d(Q^2)$ can be expressed in terms of the spectral density $\rho(\sigma)$ (4.12) by making use of Eqs. (4.2), (4.22), specifically,

$$d(Q^2) = -\frac{d}{d\ln Q^2} \int\limits_{m^2}^{\infty} \rho(\sigma) \ln\left(\frac{\sigma + Q^2}{\sigma - q_0^2} \frac{m^2 - q_0^2}{m^2 + Q^2}\right) \frac{d\sigma}{\sigma}$$

$$= Q^2 \frac{d}{dQ^2} \int\limits_{m^2}^{\infty} \ln\left(\frac{m^2 + Q^2}{\sigma + Q^2}\right) \rho(\sigma) \frac{d\sigma}{\sigma} = \frac{Q^2}{Q^2 + m^2} \int\limits_{m^2}^{\infty} \rho(\sigma) \frac{\sigma - m^2}{\sigma + Q^2} \frac{d\sigma}{\sigma}. \tag{4.23}$$

Note that the spectral density $\rho(\sigma)$ appearing in the obtained integral representations (4.18), (4.22), (4.23) can be expressed in terms of the strong corrections to the functions $\Pi(q^2)$ and $R(s)$, too. In particular, Eqs. (4.2), (4.12) imply that

$$\rho(\sigma) = \frac{1}{2\pi i} \frac{d}{d\ln\sigma} \lim_{\varepsilon \to 0_+} [p(\sigma - i\varepsilon) - p(\sigma + i\varepsilon)]. \tag{4.24}$$

In turn, relation (4.4) allows one to cast this equation to

$$\rho(\sigma) = -\frac{d}{d\ln\sigma} r(\sigma), \qquad (4.25)$$

which enables one to evaluate the spectral density in the energy intervals with available experimental data on R-ratio of electron-positron annihilation into hadrons (see also papers [336, 339, 340]).

It is also worthwhile to mention that the unified integral representations (4.18), (4.22), (4.23) for the strong corrections $p(q^2)$, $r(s)$, and $d(Q^2)$ satisfy all six relations (4.1)–(4.6) by construction. A discussion of this issue can be found in papers [337, 338]. In particular, to show that for the pair of functions [(4.22), (4.23)], the relations (4.2), (4.5) hold, one has to directly apply the differentiation and integration, respectively. The validity of Eq. (4.6) for the pair [(4.18), (4.23)] can be shown by employing the relation

$$\lim_{\varepsilon \to 0_+} \frac{1}{x \pm i\varepsilon} = \mp i\pi \delta(x) + \mathcal{P}\frac{1}{x} \qquad (4.26)$$

in the respective integrand, where $\mathcal{P}$ stands for the "Cauchy principal value." To demonstrate that the relations (4.1), (4.3) hold between pairs of expressions [(4.18), (4.22)] and [(4.18), (4.23)], the integration by parts is required. For example, Eqs. (4.3), (4.18) imply

$$d(Q^2) = Q^2 \int_{m^2}^{\infty} \frac{ds}{(s+Q^2)^2} \int_{s}^{\infty} \rho(\sigma)\frac{d\sigma}{\sigma}$$

$$= -\frac{Q^2}{s+Q^2} \int_{s}^{\infty} \rho(\sigma)\frac{d\sigma}{\sigma}\Bigg|_{s=m^2}^{s=\infty} - Q^2 \int_{m^2}^{\infty} \frac{\rho(\sigma)}{\sigma+Q^2}\frac{d\sigma}{\sigma}$$

$$= \frac{Q^2}{Q^2+m^2} \int_{m^2}^{\infty} \rho(\sigma)\frac{\sigma-m^2}{\sigma+Q^2}\frac{d\sigma}{\sigma}, \qquad (4.27)$$

which coincides with Eq. (4.23). The remaining relation (4.4) between the pair of strong corrections [(4.18), (4.22)] appears to be a bit more laborious to demonstrate than the others. Specifically, for the strong corrections $p(q^2)$ and $r(s)$ the relation (4.4) can be written as

$$r(s) = \frac{1}{2\pi i} \lim_{\varepsilon \to 0_+} \Delta p(s+i\varepsilon, s-i\varepsilon)$$

$$= \frac{1}{2\pi i} \lim_{\varepsilon \to 0_+} \left[\Delta p\left(s+i\varepsilon, q_0^2\right) - \Delta p\left(s-i\varepsilon, q_0^2\right) \right]. \qquad (4.28)$$

By virtue of Eq. (4.22)

$$\Delta p(s \pm i\varepsilon, q_0^2) = \int_{m^2}^{\infty} \rho(\sigma) \left[\ln\left(\frac{s - \sigma \pm i\varepsilon}{s - m^2 \pm i\varepsilon} \right) + \ln\left(\frac{m^2 - q_0^2}{\sigma - q_0^2} \right) \right] \frac{d\sigma}{\sigma}. \quad (4.29)$$

Then since

$$\lim_{\varepsilon \to 0_+} \ln(x \pm i\varepsilon) = \ln|x| \pm i\pi \theta(-x), \quad (4.30)$$

the first term in the square brackets on the right-hand side of Eq. (4.29) can eventually be represented as

$$\lim_{\varepsilon \to 0_+} \ln\left(\frac{s - \sigma \pm i\varepsilon}{s - m^2 \pm i\varepsilon} \right) = \ln\left| \frac{s - \sigma}{s - m^2} \right| \pm i\pi \theta(s - m^2)\theta(\sigma - s). \quad (4.31)$$

Hence, Eq. (4.29) acquires the form

$$\lim_{\varepsilon \to 0_+} \Delta p(s \pm i\varepsilon, q_0^2) = \int_{m^2}^{\infty} \rho(\sigma) \ln\left(\left| \frac{s - \sigma}{s - m^2} \right| \frac{m^2 - q_0^2}{\sigma - q_0^2} \right) \frac{d\sigma}{\sigma}$$

$$\pm i\pi \theta(s - m^2) \int_{s}^{\infty} \rho(\sigma) \frac{d\sigma}{\sigma}. \quad (4.32)$$

Therefore Eq. (4.28) reads

$$r(s) = \theta(s - m^2) \int_{s}^{\infty} \rho(\sigma) \frac{d\sigma}{\sigma}, \quad (4.33)$$

that coincides with the integral representation (4.18).

Thus the unified integral representations for the hadronic vacuum polarization function $\Pi(q^2)$, the function $R(s)$, and the Adler function $D(Q^2)$ take the following form [336–338]

$$\Delta\Pi(q^2, q_0^2) = \Delta\Pi^{(0)}(q^2, q_0^2) + \int_{m^2}^{\infty} \rho(\sigma) \ln\left(\frac{\sigma - q^2}{\sigma - q_0^2} \frac{m^2 - q_0^2}{m^2 - q^2} \right) \frac{d\sigma}{\sigma}, \quad (4.34)$$

$$R(s) = R^{(0)}(s) + \theta(s - m^2) \int_{s}^{\infty} \rho(\sigma) \frac{d\sigma}{\sigma}, \quad (4.35)$$

$$D(Q^2) = D^{(0)}(Q^2) + \frac{Q^2}{Q^2 + m^2} \int_{m^2}^{\infty} \rho(\sigma) \frac{\sigma - m^2}{\sigma + Q^2} \frac{d\sigma}{\sigma}. \quad (4.36)$$

In these equations the spectral density $\rho(\sigma)$ can be determined as the logarithmic derivative of discontinuity of the strong correction to the hadronic vacuum polarization function across the physical cut, or as the logarithmic derivative of the strong correction to the function $R(s)$, or as the discontinuity of the strong correction to the Adler function across the physical cut:

$$\rho(\sigma) = \frac{1}{2\pi i} \frac{d}{d \ln \sigma} \lim_{\varepsilon \to 0_+} [p(\sigma - i\varepsilon) - p(\sigma + i\varepsilon)] \qquad (4.37a)$$

$$= -\frac{d}{d \ln \sigma} r(\sigma) \qquad (4.37b)$$

$$= \frac{1}{2\pi i} \lim_{\varepsilon \to 0_+} [d(-\sigma - i\varepsilon) - d(-\sigma + i\varepsilon)]. \qquad (4.37c)$$

Note that for the functions $p(q^2)$ and $d(Q^2)$ satisfying conditions $p(\xi^*) = p^*(\xi)$ and $d(\xi^*) = d^*(\xi)$, Eq. (4.37) can be reduced to

$$\rho(\sigma) = \frac{1}{\pi} \frac{d}{d \ln \sigma} \text{Im} \lim_{\varepsilon \to 0_+} p(\sigma - i\varepsilon) = -\frac{dr(\sigma)}{d \ln \sigma} = \frac{1}{\pi} \text{Im} \lim_{\varepsilon \to 0_+} d(-\sigma - i\varepsilon).$$
$$(4.38)$$

The integral representations (4.34)–(4.36) constitute the "dispersively improved perturbation theory" (DPT) expressions for the functions on hand, which by construction embody all the nonperturbative constraints discussed in the beginning of this section. These representations enable one to obviate such intrinsic difficulties of the perturbative approach as the infrared unphysical singularities and provide the applicability range significantly extended with respect to that of the latter. Additionally, as it will be argued in Section 4.2 and demonstrated in Chapter 5, Eqs. (4.34)–(4.36) possess an enhanced stability[5] to the higher loop corrections as well as to the choice of the subtraction scheme at low and moderate energies. It is worthwhile to outline that the representations (4.34)–(4.36) have been obtained only from the basic dispersion relations (4.1)–(4.6) and the asymptotic ultraviolet behavior of the hadronic vacuum polarization function. Neither additional approximations nor phenomenological assumptions are required for the derivation of the unified integral representations (4.34)–(4.36).

For the nonvanishing values of m the leading-order terms in Eqs. (4.34)–(4.36) have the following form [99, 341]:

$$\Delta \Pi^{(0)}(q^2, q_0^2) = 2 \frac{\varphi - \tan \varphi}{\tan^3 \varphi} - 2 \frac{\varphi_0 - \tan \varphi_0}{\tan^3 \varphi_0}, \qquad (4.39)$$

$$R^{(0)}(s) = \theta(s - m^2) \left(1 - \frac{m^2}{s}\right)^{3/2}, \qquad (4.40)$$

$$D^{(0)}(Q^2) = 1 + \frac{3}{\xi} \left[1 - \sqrt{1 + \xi^{-1}} \, \text{arcsinh}\left(\xi^{1/2}\right)\right], \qquad (4.41)$$

where $\sin^2 \varphi = q^2/m^2$, $\sin^2 \varphi_0 = q_0^2/m^2$, and $\xi = Q^2/m^2$ (see also papers [337, 338] and references therein for the details). It is worthwhile to mention

---

5. With respect to that of the perturbative approach.

that a rather rough approximation of the leading-order terms of the functions on hand (the so-called "abrupt kinematic threshold")

$$\Delta\Pi_{AKT}^{(0)}(q^2, q_0^2) = -\ln\left(\frac{m^2 - q^2}{m^2 - q_0^2}\right), \tag{4.42}$$

$$R_{AKT}^{(0)}(s) = \theta(s - m^2), \tag{4.43}$$

$$D_{AKT}^{(0)}(Q^2) = \frac{Q^2}{Q^2 + m^2}, \tag{4.44}$$

which, nonetheless, grasps the basic peculiarities prescribed by the respective kinematic restrictions, was discussed[6] in papers [336, 339, 340, 343, 344].

It is quite straightforward to verify that the dispersion relations (4.1)–(4.6) hold for the functions (4.39)–(4.41) as well as for their approximations (4.42)–(4.44). As an example, let us demonstrate that the leading-order terms (4.43) and (4.44) satisfy the relations (4.6) and (4.3). The latter one is rather simple, since the expression (4.44) directly follows form Eqs. (4.3), (4.43), specifically,

$$D_{AKT}^{(0)}(Q^2) = Q^2 \int_{m^2}^{\infty} \frac{R_{AKT}^{(0)}(s)}{(s + Q^2)^2} ds = Q^2 \int_{m^2}^{\infty} \frac{\theta(s - m^2)}{(s + Q^2)^2} ds = \frac{Q^2}{Q^2 + m^2}. \tag{4.45}$$

The inverse relation (4.6) between the functions (4.43), (4.44) also entails no particular difficulties, namely,

$$R_{AKT}^{(0)}(s) = \frac{1}{2\pi i} \lim_{\varepsilon \to 0_+} \int_{s+i\varepsilon}^{s-i\varepsilon} D_{AKT}^{(0)}(-\zeta)\frac{d\zeta}{\zeta} = \frac{1}{2\pi i} \lim_{\varepsilon \to 0_+} \int_{s+i\varepsilon}^{s-i\varepsilon} \frac{\zeta}{\zeta - m^2}\frac{d\zeta}{\zeta}$$

$$= \frac{1}{2\pi i} \lim_{\varepsilon \to 0_+} \int_{s+i\varepsilon}^{s-i\varepsilon} \frac{d\zeta}{\zeta - m^2}. \tag{4.46}$$

For the energies exceeding the hadronic production threshold, that is, for $s \geq m^2$, in this equation it is convenient to employ the integration contour shown in Fig. 4.1A, which results in

$$R_{AKT}^{(0)}(s) = \frac{1}{2\pi i} \lim_{\varepsilon \to 0_+} \left[\int_{s+i\varepsilon}^{\infty+i\varepsilon} \frac{d\zeta}{\zeta - m^2} + \int_{C_\infty} \frac{d\zeta}{\zeta - m^2} + \int_{\infty-i\varepsilon}^{s-i\varepsilon} \frac{d\zeta}{\zeta - m^2} \right]. \tag{4.47}$$

Because the integrand on the right-hand side of Eq. (4.46) contains no singularities in the complex ζ-plane except for the pole of first order at $\zeta = m^2$, the first

---

6. The approximations (4.42)–(4.44) (with m being an adjustable parameter) have also been discussed in paper [342].

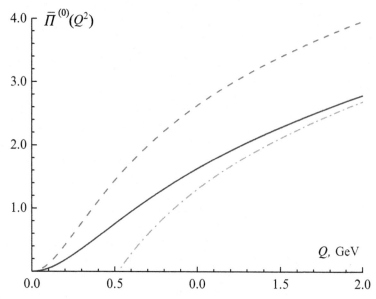

FIG. 4.2 The leading-order term of the subtracted hadronic vacuum polarization function $\bar{\Pi}(Q^2) = \Delta\Pi(0, -Q^2)$ (Eq. 4.39, *solid curve*), its approximate form (Eq. 4.42, *dashed curve*), and the corresponding perturbative expression (Eq. 3.31, *dot-dashed curve*).

and third terms on the right-hand side of Eq. (4.47) cancel each other. It is worth noting here that the integration of the leading-order term of the Adler function along the subcontour c_∞ in Eq. (4.47) gives a nonvanishing[7] contribution to the function $R(s)$. Specifically, in the remaining second term of Eq. (4.47) it is convenient to change the integration variable $\zeta = m^2 + r\,e^{i\varphi}$, which casts it to

$$R_{\text{AKT}}^{(0)}(s) = \frac{1}{2\pi i}\lim_{\varepsilon \to 0_+}\lim_{r \to \infty}\int_{\varepsilon}^{2\pi-\varepsilon}\frac{r\,i e^{i\varphi}}{r\,e^{i\varphi}}d\varphi = 1, \quad s \geq m^2. \tag{4.48}$$

Following the very same lines as above, one may also demonstrate that for $s \geq m^2$ the integration in Eq. (4.46) along the contour shown in Fig. 4.1B yields the result identical to Eq. (4.48). At the same time it is easy to show that for the energies below the hadronic production threshold ($s < m^2$) the result of integration in Eq. (4.46) along either of the contours shown in Fig. 4.1C and D vanishes, namely,

$$R_{\text{AKT}}^{(0)}(s) = 0, \quad s < m^2. \tag{4.49}$$

---

7. Contrary to the respective contribution to the function $R(s)$ given by the strong correction to the Adler function, see Eq. (4.10).

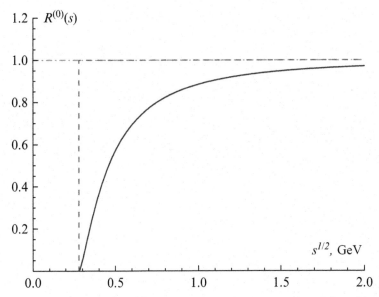

FIG. 4.3 The leading-order term of the function $R(s)$ (Eq. 4.40, *solid curve*), its approximate form (Eq. 4.43, *dashed line*), and the corresponding perturbative expression (Eq. 3.61, *dot-dashed line*).

Hence, Eqs. (4.48), (4.49) imply

$$R_{AKT}^{(0)}(s) = \frac{1}{2\pi i} \lim_{\varepsilon \to 0_+} \int_{s+i\varepsilon}^{s-i\varepsilon} \frac{d\zeta}{\zeta - m^2} = \theta(s - m^2), \qquad (4.50)$$

which coincides with the expression (4.43). This example, in particular, illustrates the fact that the absence of the unphysical singularities in the leading-order term of the Adler function (4.44) allows one to perform in Eq. (4.46) the continuous deformation of the integration contours shown in Fig. 4.1A and C into the corresponding contours shown in Fig. 4.1B and D and vice versa.

As another example, it is also worthwhile to demonstrate that the leading-order terms (4.39), (4.40) satisfy the relation (4.4). The calculation of the discontinuity (4.4) of the function $\Pi(s)$ yields a nonvanishing result only on its physical cut, namely, for the energies exceeding the threshold ($s \geq m^2$). In this case

$$\lim_{\varepsilon \to 0_+} \arcsin\left(\sqrt{x \pm i\varepsilon}\right) = \frac{\pi}{2} \pm i\arcsinh\left(\sqrt{x-1}\right), \qquad (4.51)$$

$$\lim_{\varepsilon \to 0_+} \tan\left[\arcsin\left(\sqrt{x \pm i\varepsilon}\right)\right] = \pm i\coth\left[\arcsinh\left(\sqrt{x-1}\right)\right] = \pm i\sqrt{\frac{x}{x-1}}, \qquad (4.52)$$

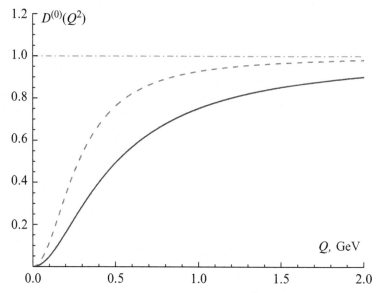

FIG. 4.4 The leading-order term of the Adler function $D(Q^2)$ (Eq. 4.41, *solid curve*), its approximate form (Eq. 4.44, *dashed curve*), and the corresponding perturbative expression (Eq. 3.53, *dot-dashed line*).

and

$$\lim_{\varepsilon \to 0_+} \Pi^{(0)}(s \pm i\varepsilon) = \left(1 - \frac{1}{x}\right)^{3/2} \left[2\sqrt{\frac{x}{x-1}} - 2\operatorname{arcsinh}\left(\sqrt{x-1}\right) \pm i\pi\right], \tag{4.53}$$

where $x = s/m^2$. Therefore the discontinuity (4.4) of the function (4.39) takes the form

$$R^{(0)}(s) = \frac{1}{2\pi i} \lim_{\varepsilon \to 0_+} \Delta \Pi^{(0)}(s + i\varepsilon, s - i\varepsilon) = \theta(s - m^2)\left(1 - \frac{m^2}{s}\right)^{3/2}, \tag{4.54}$$

which is identical to Eq. (4.40).

The plots of the leading-order terms of the functions on hand $\Pi^{(0)}(q^2)$, $R^{(0)}(s)$, and $D^{(0)}(Q^2)$ are shown in Figs. 4.2–4.4, respectively. As one can infer from these figures, the expressions (4.39)–(4.41), as well as their approximations (4.42)–(4.44), tend to their perturbative predictions (3.31), (3.61), and (3.53) at high energies. At the same time, Figs. 4.2–4.4 clearly illustrate that the functions (4.39)–(4.41), as well as (4.42)–(4.44), comply with the aforementioned low-energy kinematic restrictions, whereas their perturbative approximations (3.31), (3.61), (3.53) are incompatible with the latter. The inclusion of the higher-order terms of the functions on hand requires the calculation of the relevant spectral density $\rho(\sigma)$ (4.37), which will be addressed in the next section.

4.2   SPECTRAL DENSITY

In general, a rigorous method of theoretical description of the hadronic vacuum polarization function $\Pi(q^2)$, the function $R(s)$, and the Adler function $D(Q^2)$, which would have provided one with robust unabridged results applicable in the entire energy range, is still far from being feasible. In particular, this fact implies that for the spectral density $\rho(\sigma)$ (4.37) entering the unified integral representations for the functions on hand (4.34)–(4.36) only approximate expression is available so far, too. For example, the definition (4.37b), which enables one to extract the numerical values of $\rho(\sigma)$ from the experimental data on R-ratio of electron-positron annihilation into hadrons, is applicable only in the energy intervals where the corresponding measurements have been performed. In turn, the definitions of the spectral density (4.37a), (4.37c) require an explicit theoretical expression for the strong correction to the hadronic vacuum polarization function, $p(q^2)$, and to the Adler function, $d(Q^2)$, respectively. The QCD perturbation theory still remains the basic method of calculation of these functions. However, the perturbative approximations of $p(q^2)$ and $d(Q^2)$ are applicable at moderate and high energies only, since they contain infrared unphysical singularities. The latter, being an inherent difficulty of the perturbative approach to QCD, make such approximations of the functions on hand incompatible with corresponding dispersion relations and eventually force one to invoke the nonperturbative methods for the study of hadron dynamics at low energies.

As mentioned in the previous section, the spectral density $\rho(\sigma)$ provides the corresponding perturbative input to the unified integral representations (4.34)–(4.36). In particular, the perturbative part of the spectral density can be calculated by making use of the strong correction to either of the functions on hand, namely,

$$\rho_{\text{pert}}(\sigma) = \frac{1}{2\pi i} \frac{d}{d\ln\sigma} \lim_{\varepsilon \to 0_+} \left[p_{\text{pert}}(\sigma - i\varepsilon) - p_{\text{pert}}(\sigma + i\varepsilon) \right] \qquad (4.55a)$$

$$= -\frac{d}{d\ln\sigma} r_{\text{pert}}(\sigma) \qquad (4.55b)$$

$$= \frac{1}{2\pi i} \lim_{\varepsilon \to 0_+} \left[d_{\text{pert}}(-\sigma - i\varepsilon) - d_{\text{pert}}(-\sigma + i\varepsilon) \right]. \qquad (4.55c)$$

In Eq. (4.55b) it is assumed that the strong correction $r_{\text{pert}}(s)$ thoroughly accounts for the effects due to the continuation of the spacelike theoretical results into the timelike domain. It is worthwhile to note here that the spectral function $\rho_{\text{pert}}(\sigma)$ (4.55) embodies only perturbative contributions and, in general, its behavior at small values of σ may be altered by the terms of an intrinsically nonperturbative nature. For example, the models for the spectral density proposed in papers [337, 345, 346] constitute a superposition of the perturbative part (4.55) with the so-called "flat" terms, which by definition do not contribute to the corresponding perturbative expressions at high energies. At the same time, there are also a number of models that modify the infrared

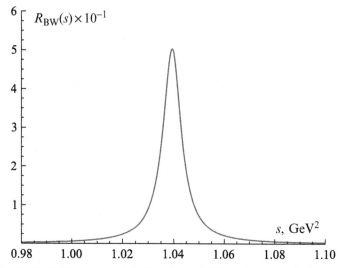

FIG. 4.5 The Breit-Wigner parameterization of the $\phi(1020)$ resonance state (4.56).

behavior of $\rho_{\text{pert}}(\sigma)$ proceeding from a certain phenomenological reasoning (see, e.g., papers [347, 348]). Nevertheless, the qualitative behavior of the functions on hand (4.34)–(4.36) in the entire energy range proves to be basically independent[8] of a particular model for the spectral function. Specifically, the ultraviolet asymptotic of the functions (4.34)–(4.36) is fixed by the perturbation theory, whereas their behavior at low energies is bound by the nonperturbative constraints discussed in the previous section (see also Chapter 5).

It is also worthwhile to note that, as discussed earlier, the theoretical expression for the R-ratio of electron-positron annihilation into hadrons is the result of continuation of the Adler function $D(Q^2)$ (4.2) into the timelike domain (4.6). Undoubtedly, this procedure is well defined and yields a unique result for the function $R(s)$, if a comprehensive unabridged expression for the Adler function is employed. However, since the latter is not yet attainable, in practice one inevitably resorts to an approximate form of $D(Q^2)$, such as its perturbative expression (3.57). At the same time one has to be aware that the result of continuation of the perturbative approximation of the Adler function (3.57) into the timelike domain, in general, depends on the choice of the pertinent integration contour in Eq. (4.6). Basically, this is caused by the fact that the analytic properties of the Adler function in the kinematic variable Q^2 prescribed by the dispersion relation (4.3) differ from those of its perturbative approximation (3.57). In particular, the latter possesses the infrared unphysical singularities at positive values of real Q^2, that, in turn, does not allow one to

---

8. Contrary to the behavior of these functions in the massless limit (4.107)–(4.109).

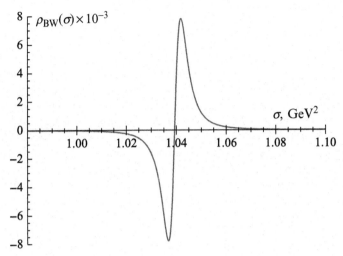

FIG. 4.6 The spectral function $\rho_{BW}(\sigma)$ (4.57) corresponding to the Breit-Wigner parameterization of the $\phi(1020)$ resonance state.

continuously deform the respective integration contours shown in Fig. 4.1 into each other. Nonetheless, the use of the pair of integration contours shown in Fig. 4.1A and C in the calculation of the function $R(s)$ (4.6) eventually yields a physically meaningful result,[9] even if the perturbative approximation of the Adler function (3.57) is employed for the energies above the threshold.

Basically, in addition to the perturbative part of the spectral density (4.55), one may also take into account the contributions describing the resonance structure of the R-ratio of electron-positron annihilation into hadrons. For this purpose it is convenient to employ the Breit-Wigner parameterization, which for the case of a narrow resonance reads (see, e.g., paper [349])

$$R_{BW}(s) = \frac{9}{\alpha_{em}^2(M^2)} \frac{\Gamma\Gamma_{ee}M^2}{(s - M^2)^2 + \Gamma^2 M^2}. \tag{4.56}$$

In this equation M is the mass of a resonance state, whereas Γ and Γ_{ee} denote its full width and its partial width for the decay into the electron-positron channel, respectively. As an example, the plot of the function $R_{BW}(s)$ (4.56) corresponding to the $\phi(1020)$ resonance is shown in Fig. 4.5, the values of the respective parameters (specifically, $M \simeq 1.019\,\text{GeV}$, $\Gamma \simeq 4.266\,\text{MeV}$, and $\Gamma_{ee} \simeq 1.27\,\text{keV}$) being taken from review [84]. In turn, the corresponding contribution to the spectral density (4.37b) acquires the following form

---

9. Namely, the function $R(s)$ assumes nonzero values only for the energies exceeding the hadronic production threshold and its high-energy behavior complies with relevant experimental observations.

$$\rho_{\text{BW}}(\sigma) = \frac{18}{\alpha_{\text{em}}^2(M^2)} \frac{\Gamma\Gamma_{\text{ee}}M^2\sigma(\sigma - M^2)}{\left[(\sigma - M^2)^2 + \Gamma^2 M^2\right]^2}. \tag{4.57}$$

Note that in Eqs. (4.56), (4.57) the common prefactor (1.32) is effectively accounted for. The spectral function (4.57), which corresponds to the $\phi(1020)$ resonance state, is shown in Fig. 4.6. As one can infer from this figure the Breit-Wigner parameterization of a resonance contribution to the spectral density (4.57) is the function of an alternating sign, specifically, $\rho_{\text{BW}}(\sigma) < 0$ for $\sigma < M^2$, $\rho_{\text{BW}}(M^2) = 0$, and $\rho_{\text{BW}}(\sigma) > 0$ for $\sigma > M^2$. It is also worthwhile to mention that in practical applications, instead of parameterizing the resonance structure of the function $R(s)$, one may alternatively use the widely employed method of "smearing" of the R-ratio of electron-positron annihilation into hadrons [102], which will be discussed in Section 5.2.

Let us address now the calculation of the perturbative spectral function $\rho_{\text{pert}}(\sigma)$. For this purpose it is most convenient to employ the definition (4.55c)

$$\rho_{\text{pert}}^{(\ell)}(\sigma) = \frac{1}{2\pi i} \lim_{\varepsilon \to 0_+} \left[d_{\text{pert}}^{(\ell)}(-\sigma - i\varepsilon) - d_{\text{pert}}^{(\ell)}(-\sigma + i\varepsilon) \right], \tag{4.58}$$

which involves the strong correction to the Adler function $d_{\text{pert}}^{(\ell)}(Q^2)$. As noted earlier, at the ℓ-loop level the latter can be represented as the power series in the perturbative QCD "couplant" $a_{\text{s}}^{(\ell)}(Q^2) = \alpha_{\text{s}}^{(\ell)}(Q^2)\beta_0/(4\pi)$, specifically (3.57),

$$d_{\text{pert}}^{(\ell)}(Q^2) = \sum_{j=1}^{\ell} d_j \left[a_{\text{s}}^{(\ell)}(Q^2) \right]^j. \tag{4.59}$$

The available perturbative expansion coefficients of the Adler function d_j ($1 \le j \le 4$) can be found in Section 3.2, whereas the explicit expressions for $a_{\text{s}}^{(\ell)}(Q^2)$ are given in Chapter 2 and Appendix A.

At the one-loop level ($\ell = 1$) the right-hand side of Eq. (4.59) consists of only one term

$$d_{\text{pert}}^{(1)}(Q^2) = d_1 \frac{1}{\ln(Q^2/\Lambda^2)}, \tag{4.60}$$

which makes the calculation of the corresponding perturbative spectral function quite straightforward. In Eq. (4.60) $d_1 = 4/\beta_0$ and $\beta_0 = 11 - 2n_f/3$ stand for the one-loop perturbative expansion coefficients of the Adler function and β function, respectively, n_f is the number of active flavors, and Λ denotes the QCD scale parameter. Hence, the one-loop perturbative spectral function (4.58) reads

$$\rho_{\text{pert}}^{(1)}(\sigma) = d_1 \frac{1}{2\pi i} \lim_{\varepsilon \to 0_+} \left\{ \frac{1}{\ln\left[-(\sigma/\Lambda^2) - i\varepsilon\right]} - \frac{1}{\ln\left[-(\sigma/\Lambda^2) + i\varepsilon\right]} \right\}. \tag{4.61}$$

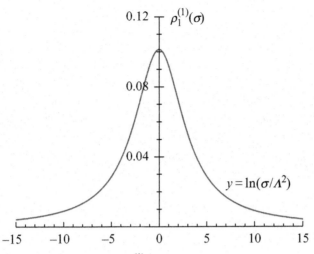

FIG. 4.7 The one-loop spectral function $\varrho_1^{(1)}(\sigma)$ (4.63).

Then by virtue of Eq. (4.30) the function $\rho_{\text{pert}}^{(1)}(\sigma)$ for the positive values of its argument acquires the following form

$$\rho_{\text{pert}}^{(1)}(\sigma) = d_1 \frac{1}{2\pi i} \left[\frac{1}{\ln(\sigma/\Lambda^2) - i\pi} - \frac{1}{\ln(\sigma/\Lambda^2) + i\pi} \right], \qquad (4.62)$$

which can eventually be represented as

$$\rho_{\text{pert}}^{(1)}(\sigma) = d_1 \varrho_1^{(1)}(\sigma), \quad \varrho_1^{(1)}(\sigma) = \frac{1}{y^2 + \pi^2}, \quad y = \ln\left(\frac{\sigma}{\Lambda^2}\right). \qquad (4.63)$$

The one-loop spectral function $\varrho_1^{(1)}(\sigma)$ (4.63), being expressed in terms of the dimensionless variable $y = \ln(\sigma/\Lambda^2)$, is an even function of y. The function (4.63) assumes the values in the interval $0 \le \varrho_1^{(1)}(\sigma) \le 1/\pi^2$ and decreases as $1/y^2$ in both ultraviolet ($y \to \infty$) and infrared ($y \to -\infty$) asymptotics. The plot of the spectral function $\varrho_1^{(1)}(\sigma)$ (4.63) is shown in Fig. 4.7.

At the higher loop levels ($\ell \ge 2$) the perturbative spectral function (4.58) can be represented as

$$\rho_{\text{pert}}^{(\ell)}(\sigma) = \sum_{j=1}^{\ell} d_j \varrho_j^{(\ell)}(\sigma), \qquad (4.64)$$

where $\varrho_j^{(\ell)}(\sigma)$ stands for the "partial" spectral function corresponding to the jth power of the ℓ-loop perturbative couplant:

$$\varrho_j^{(\ell)}(\sigma) = \frac{1}{2\pi i} \lim_{\varepsilon \to 0_+} \left\{ \left[a_s^{(\ell)}(-\sigma - i\varepsilon) \right]^j - \left[a_s^{(\ell)}(-\sigma + i\varepsilon) \right]^j \right\}. \qquad (4.65)$$

The explicit calculation of $\rho_{\text{pert}}^{(\ell)}(\sigma)$ (4.64) is somewhat impeded. In particular, since the expression for the strong running coupling $\alpha_{\text{s}}^{(\ell)}(Q^2)$ becomes rather cumbrous at the higher loop levels, the calculation of its discontinuity across the physical cut, $\varrho_1^{(\ell)}(\sigma)$, as well as the calculation of the discontinuity of its integer powers, $\varrho_j^{(\ell)}(\sigma)$ $(2 \leq j \leq \ell)$, entails certain technical difficulties. Nonetheless, the latter can be circumvented by the following quite simple method, which enables one to explicitly calculate the partial spectral function $\varrho_j^{(\ell)}(\sigma)$ (4.65) corresponding to any integer power j of the strong running coupling $\alpha_{\text{s}}^{(\ell)}(Q^2)$, and thereby to construct the whole perturbative spectral function $\rho_{\text{pert}}^{(\ell)}(\sigma)$ (4.64) at an arbitrary[10] loop level (see also papers [350–352]).

First of all, to obtain the perturbative spectral function $\varrho_j^{(\ell)}(\sigma)$ (4.65) for any $j \geq 1$ it appears to be enough to calculate only real and imaginary parts of the ℓ-loop couplant $a_{\text{s}}^{(\ell)}(Q^2)$ at the edges of its cut:

$$\lim_{\varepsilon \to 0_+} a_{\text{s}}^{(\ell)}(-\sigma \pm i\varepsilon) = a_{\text{Re}}^{(\ell)}(\sigma) \mp i\pi a_{\text{Im}}^{(\ell)}(\sigma), \qquad (4.66)$$

where $a_{\text{Re}}^{(\ell)}(\sigma)$ and $a_{\text{Im}}^{(\ell)}(\sigma)$ are the real functions of their arguments and $\sigma \geq 0$ is assumed. It is obvious that

$$a_{\text{Re}}^{(\ell)}(\sigma) = \frac{1}{2} \lim_{\varepsilon \to 0_+} \left[a_{\text{s}}^{(\ell)}(-\sigma - i\varepsilon) + a_{\text{s}}^{(\ell)}(-\sigma + i\varepsilon) \right], \qquad (4.67)$$

and

$$a_{\text{Im}}^{(\ell)}(\sigma) = \frac{1}{2\pi i} \lim_{\varepsilon \to 0_+} \left[a_{\text{s}}^{(\ell)}(-\sigma - i\varepsilon) - a_{\text{s}}^{(\ell)}(-\sigma + i\varepsilon) \right]. \qquad (4.68)$$

In this case, Eq. (4.66) implies that for integer nonnegative values of j the following equation holds

$$\lim_{\varepsilon \to 0_+} \left[a_{\text{s}}^{(\ell)}(-\sigma \pm i\varepsilon) \right]^j = \sum_{k=0}^{j} \binom{j}{k} (\mp i\pi)^k \left[a_{\text{Re}}^{(\ell)}(\sigma) \right]^{j-k} \left[a_{\text{Im}}^{(\ell)}(\sigma) \right]^k, \qquad (4.69)$$

where

$$\binom{n}{m} = \frac{n!}{m!(n-m)!} \qquad (4.70)$$

---

10. It is assumed that the involved perturbative coefficients d_j and β_j are known.

denotes the binomial coefficient. In turn, for $j \geq 1$ Eq. (4.69) can be represented as

$$
\lim_{\varepsilon \to 0_+} \left[a_{\mathrm{s}}^{(\ell)}(-\sigma \pm i\varepsilon) \right]^j = \sum_{k=0}^{K(j+1)} \binom{j}{2k} (-1)^k (\pi)^{2k} \left[a_{\mathrm{Re}}^{(\ell)}(\sigma) \right]^{j-2k} \left[a_{\mathrm{Im}}^{(\ell)}(\sigma) \right]^{2k}
$$

$$
\mp i\pi \sum_{k=0}^{K(j)} \binom{j}{2k+1} (-1)^k (\pi)^{2k} \left[a_{\mathrm{Re}}^{(\ell)}(\sigma) \right]^{j-2k-1} \left[a_{\mathrm{Im}}^{(\ell)}(\sigma) \right]^{2k+1}, \quad (4.71)
$$

where

$$
K(j) = \frac{j}{2} + \frac{j \bmod 2}{2} - 1 \tag{4.72}
$$

and $(j \bmod n)$ is the remainder on division of j by n, that is, $(j \bmod 2) = 0$ for even values of j and $(j \bmod 2) = 1$ for odd values of j. The functions $K(j+1)$ and $K(j)$ specify[11] the numbers of terms on the right-hand side of Eq. (4.69), which correspond to the even and odd values of k, respectively. Therefore Eq. (4.65) reads

$$
\varrho_j^{(\ell)}(\sigma) = \sum_{k=0}^{K(j)} \binom{j}{2k+1} (-1)^k \pi^{2k} \left[a_{\mathrm{Re}}^{(\ell)}(\sigma) \right]^{j-2k-1} \left[a_{\mathrm{Im}}^{(\ell)}(\sigma) \right]^{2k+1}, \tag{4.73}
$$

with $\sigma \geq 0$ and $j \geq 1$ being assumed. In particular, the first four relations (4.73) acquire a simple form

$$
\varrho_1^{(\ell)}(\sigma) = a_{\mathrm{Im}}^{(\ell)}(\sigma), \tag{4.74}
$$

$$
\varrho_2^{(\ell)}(\sigma) = 2a_{\mathrm{Im}}^{(\ell)}(\sigma) a_{\mathrm{Re}}^{(\ell)}(\sigma), \tag{4.75}
$$

$$
\varrho_3^{(\ell)}(\sigma) = a_{\mathrm{Im}}^{(\ell)}(\sigma) \left\{ 3 \left[a_{\mathrm{Re}}^{(\ell)}(\sigma) \right]^2 - \pi^2 \left[a_{\mathrm{Im}}^{(\ell)}(\sigma) \right]^2 \right\}, \tag{4.76}
$$

$$
\varrho_4^{(\ell)}(\sigma) = 4a_{\mathrm{Im}}^{(\ell)}(\sigma) a_{\mathrm{Re}}^{(\ell)}(\sigma) \left\{ \left[a_{\mathrm{Re}}^{(\ell)}(\sigma) \right]^2 - \pi^2 \left[a_{\mathrm{Im}}^{(\ell)}(\sigma) \right]^2 \right\}. \tag{4.77}
$$

Then, the functions $a_{\mathrm{Re}}^{(\ell)}(\sigma)$ and $a_{\mathrm{Im}}^{(\ell)}(\sigma)$ entering Eq. (4.73) can be calculated explicitly in a quite similar way. Specifically, as discussed in Section 2.3, the perturbative QCD couplant $a_{\mathrm{s}}^{(\ell)}(Q^2)$ can be represented as the double sum (2.83)

$$
a_{\mathrm{s}}^{(\ell)}(Q^2) = \sum_{n=1}^{\ell} \sum_{m=0}^{n-1} b_n^m \bar{a}_n^m(Q^2), \tag{4.78}
$$

where

$$
\bar{a}_n^m(Q^2) = \frac{\ln^m(\ln z)}{\ln^n z}, \quad z = \frac{Q^2}{\Lambda^2}. \tag{4.79}
$$

---

11. Note that the function $K(j)$ (4.72) is applicable only for positive values of its argument.

Recall that for $\ell \geq 2$ Eq. (4.78) constitutes the iterative expression for the perturbative QCD couplant, the difference between Eq. (4.78) and the corresponding exact solution to the renormalization group equation (2.12) being uncontrollable at any given loop level. The numerical coefficients b_n^m entering Eq. (4.78) are the combinations of the β function perturbative expansion coefficients, in particular, $b_1^0 = 1, b_2^0 = 0, b_2^1 = -B_1 = -\beta_1/\beta_0^2$, see Appendix A. For practical purposes it is convenient to express the functions $a_{\mathrm{Re}}^{(\ell)}(\sigma)$ and $a_{\mathrm{Im}}^{(\ell)}(\sigma)$ (4.66) in terms of real and imaginary parts of $\bar{a}_n^m(Q^2)$ (4.79) at the edges of its cut, namely,

$$a_{\mathrm{Re}}^{(\ell)}(\sigma) = \sum_{n=1}^{\ell} \sum_{m=0}^{n-1} b_n^m u_n^m(\sigma), \tag{4.80}$$

$$a_{\mathrm{Im}}^{(\ell)}(\sigma) = \sum_{n=1}^{\ell} \sum_{m=0}^{n-1} b_n^m v_n^m(\sigma), \tag{4.81}$$

where

$$\lim_{\varepsilon \to 0_+} \bar{a}_n^m(-\sigma \pm i\varepsilon) = u_n^m(\sigma) \mp i\pi v_n^m(\sigma), \tag{4.82}$$

$u_n^m(\sigma)$ and $v_n^m(\sigma)$ are the real functions of their arguments, and $\sigma \geq 0$ is assumed. It is quite evident that

$$u_n^m(\sigma) = \frac{1}{2} \lim_{\varepsilon \to 0_+} \left[\bar{a}_n^m(-\sigma - i\varepsilon) + \bar{a}_n^m(-\sigma + i\varepsilon) \right] \tag{4.83}$$

and

$$v_n^m(\sigma) = \frac{1}{2\pi i} \lim_{\varepsilon \to 0_+} \left[\bar{a}_n^m(-\sigma - i\varepsilon) - \bar{a}_n^m(-\sigma + i\varepsilon) \right]. \tag{4.84}$$

On the right-hand side of Eq. (4.78) and on the left-hand side of Eq. (4.79), as well as in Eqs. (4.80)–(4.84), the integer superscripts m are not to be confused with respective powers.

To calculate the functions $u_n^m(\sigma)$ and $v_n^m(\sigma)$ entering Eq. (4.82), it is convenient to split the left-hand side of the latter into two factors:

$$\lim_{\varepsilon \to 0_+} \bar{a}_n^m(-\sigma \pm i\varepsilon) = \lim_{\varepsilon \to 0_+} \left[\bar{a}_n^0(-\sigma \pm i\varepsilon) \bar{a}_0^m(-\sigma \pm i\varepsilon) \right]. \tag{4.85}$$

Let us address the first factor on the right-hand side of this equation, which reads

$$\lim_{\varepsilon \to 0_+} \bar{a}_n^0(-\sigma \pm i\varepsilon) = \frac{(y \mp i\pi)^n}{(y^2 + \pi^2)^n}, \quad y = \ln\left(\frac{\sigma}{\Lambda^2}\right). \tag{4.86}$$

Proceeding along the same lines as in the previously studied case, one can cast the numerator on the right-hand side of this equation to

$$(y \mp i\pi)^n = \sum_{k=0}^{K(n+1)} \binom{n}{2k} (-1)^k \pi^{2k} y^{n-2k} \mp i\pi \sum_{k=0}^{K(n)} \binom{n}{2k+1} (-1)^k \pi^{2k} y^{n-2k-1}. \tag{4.87}$$

Therefore the functions (4.83), (4.84) for $m = 0$ take the form

$$u_n^0(\sigma) = \frac{1}{(y^2 + \pi^2)^n} \sum_{k=0}^{K(n+1)} \binom{n}{2k} (-1)^k \pi^{2k} y^{n-2k} \qquad (4.88)$$

and

$$v_n^0(\sigma) = \frac{1}{(y^2 + \pi^2)^n} \sum_{k=0}^{K(n)} \binom{n}{2k+1} (-1)^k \pi^{2k} y^{n-2k-1}, \qquad (4.89)$$

with $n \geq 1$ being assumed.

Let us turn now to the second factor on the right-hand side of Eq. (4.85), which takes the following form for $m \geq 1$:

$$\lim_{\varepsilon \to 0_+} \bar{a}_0^m(-\sigma \pm i\varepsilon) = [\ln(y \pm i\pi)]^m, \qquad (4.90)$$

where $y = \ln(\sigma/\Lambda^2)$. Since for real a and b

$$\ln(a \pm ib) = \ln\sqrt{a^2 + b^2} \pm i\pi \left[\frac{1}{2} - \frac{1}{\pi} \arctan\left(\frac{a}{b}\right) \right], \quad b > 0, \qquad (4.91)$$

one can cast the right-hand side of Eq. (4.90) to

$$[\ln(y \pm i\pi)]^m = \sum_{k=0}^{m} \binom{m}{k} (\pm i\pi)^k [L_1(y)]^{m-k} [L_2(y)]^k, \qquad (4.92)$$

where

$$L_1(y) = \ln\sqrt{y^2 + \pi^2} \qquad (4.93)$$

and

$$L_2(y) = \frac{1}{2} - \frac{1}{\pi} \arctan\left(\frac{y}{\pi}\right). \qquad (4.94)$$

In Eqs. (4.91), (4.94) it is assumed that $\arctan(x)$ is a monotone nondecreasing function of its argument: $-\pi/2 \leq \arctan(x) \leq \pi/2$ for $-\infty < x < \infty$. Finally following the very same lines as above, one can represent Eq. (4.92) as

$$[\ln(y \pm i\pi)]^m = u_0^m(\sigma) \mp i\pi v_0^m(\sigma), \qquad (4.95)$$

where

$$u_0^m(\sigma) = \sum_{k=0}^{K(m+1)} \binom{m}{2k} (-1)^k \pi^{2k} [L_1(y)]^{m-2k} [L_2(y)]^{2k}, \qquad (4.96)$$

$$v_0^m(\sigma) = \sum_{k=0}^{K(m)} \binom{m}{2k+1} (-1)^{k+1} \pi^{2k} [L_1(y)]^{m-2k-1} [L_2(y)]^{2k+1}, \qquad (4.97)$$

the function $K(m)$ is defined in Eq. (4.72), and $m \geq 1$ is assumed in Eq. (4.97).

Therefore the functions $u_n^m(\sigma)$ (4.83) and $v_n^m(\sigma)$ (4.84) acquire the following form:

$$u_n^m(\sigma) = \begin{cases} u_n^0(\sigma), & \text{if } m = 0, \\ u_n^0(\sigma)u_0^m(\sigma) - \pi^2 v_n^0(\sigma)v_0^m(\sigma), & \text{if } m \geq 1, \end{cases} \qquad (4.98)$$

and

$$v_n^m(\sigma) = \begin{cases} v_n^0(\sigma), & \text{if } m = 0, \\ v_n^0(\sigma)u_0^m(\sigma) + u_n^0(\sigma)v_0^m(\sigma), & \text{if } m \geq 1, \end{cases} \qquad (4.99)$$

with $n \geq 1$ being assumed. The functions appearing on the right-hand sides of these equations are defined in Eqs. (4.88), (4.89), (4.96), (4.97). Thus the ℓ-loop spectral function $\rho_{\text{pert}}^{(\ell)}(\sigma)$ (4.58) can be explicitly calculated at an arbitrary loop level by making use of the relations (4.64), (4.73), (4.80), (4.81), (4.98), (4.99), the availability of the involved perturbative coefficients d_j and β_j being assumed. The explicit expressions for the perturbative spectral functions $\rho_{\text{pert}}^{(\ell)}(\sigma)$ (4.58) up to the four-loop level ($1 \leq \ell \leq 4$) are presented in Appendix B (see also papers [350–352] and references therein).

It is interesting to note here that the leading-order term of the expansion of function $v_n^m(\sigma)$ (4.99) in the ultraviolet asymptotic $\sigma \to \infty$ turns out to be identical to the spectral function $\bar{v}_n^m(\sigma)$, which corresponds to the "naive" continuation of the spacelike theoretical results into the timelike domain (3.64), namely,

$$v_n^m(\sigma) \simeq \bar{v}_n^m(\sigma) + \mathcal{O}\left[\left(\frac{\Lambda^2}{\sigma}\right)^{n+3}\right], \qquad \sigma \to \infty, \qquad (4.100)$$

where

$$\bar{v}_n^m(\sigma) = -\frac{d}{d\ln\sigma}\bar{a}_n^m(|\sigma|) = \frac{(\ln y)^{m-1}}{y^{n+1}}(n\ln y - m), \qquad y = \ln\left(\frac{\sigma}{\Lambda^2}\right), \quad (4.101)$$

the function $\bar{a}_n^m(Q^2)$ is defined in Eq. (4.79), and the definition (4.55b) is employed. As one can infer from Fig. 4.8, the functions (4.99), (4.101) coincide at $\sigma \to \infty$, whereas their behavior at moderate and low σ is qualitatively different. In particular, $\bar{v}_n^m(\sigma)$ diverges at $\sigma \to \Lambda^2$. Obviously, the spectral function $\rho(\sigma)$ composed of the expressions $\bar{v}_n^m(\sigma)$ (4.101) is not integrable in the representations (4.34), (4.36), whereas its integrability in Eq. (4.35) depends on whether the energy s is high enough to avoid all the singularities of the integrand.

It is also worthwhile to mention that Eqs. (4.100), (4.101) signify that at an arbitrary loop level the integral representation (4.35), being reexpanded at $s \to \infty$, reproduces term-by-term the entire perturbative approximation of the Adler function, which constitutes the so-called naive continuation of $D_{\text{pert}}^{(\ell)}(Q^2)$ into the timelike domain (3.64). At the same time the reexpanded integral

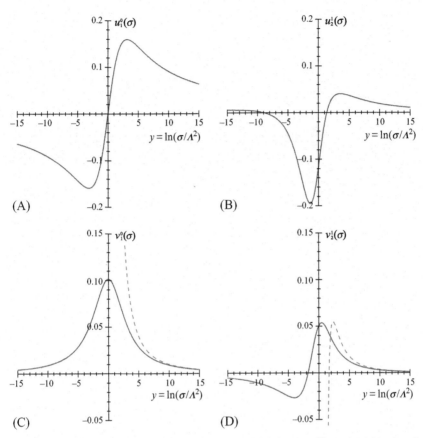

FIG. 4.8 The real ($u_n^m(\sigma)$, Eq. 4.98, *plots A and B*) and imaginary ($v_n^m(\sigma)$, Eq. 4.99, *plots C and D*) parts of the function $\bar{a}_n^m(Q^2)$ (4.79) on its cut, see Eq. (4.82). *Plots C and D also display the function $\bar{v}_n^m(\sigma)$ (4.101) corresponding to the naive continuation of $\bar{a}_n^m(Q^2)$ into the timelike domain (dashed curve). The values of parameters: $n = 1$, $m = 0$ (plots A and C); $n = 2$, $m = 1$ (plots B and D); $y = \ln(\sigma/\Lambda^2)$.*

representation (4.35) additionally produces an infinite number of terms,[12] which cannot be neglected starting from the three-loop level even in the ultraviolet asymptotic $s \to \infty$. As it will be discussed later (see Section 5.2, Chapter 6, and Appendix C), these π^2-terms appear to dominate over the corresponding perturbative contributions and thereby play a substantial role in the study of the R-ratio of electron-positron annihilation into hadrons and the related quantities in the entire energy range.

The partial spectral functions $\varrho_j^{(\ell)}(\sigma)$ (4.65), which correspond to the jth power ($1 \le j \le \ell$) of the ℓ-loop ($1 \le \ell \le 4$) perturbative QCD

---

12. Such terms are proportional to integer powers of π^2, so that they are called "π^2-terms."

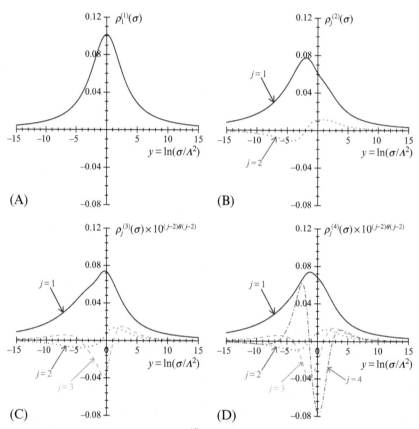

FIG. 4.9 The partial spectral functions $\varrho_j^{(\ell)}(\sigma)$ (4.65) corresponding to jth power ($1 \le j \le \ell$) of the ℓ-loop ($1 \le \ell \le 4$) perturbative QCD couplant $a_s^{(\ell)}(Q^2)$. *Plot A*: one-loop level ($\ell = 1$, $j = 1$). *Plot B*: two-loop level ($\ell = 2$, $1 \le j \le 2$). *Plot C*: three-loop level ($\ell = 3$, $1 \le j \le 3$), the function $\varrho_3^{(3)}(\sigma)$ is scaled by the factor of 10. *Plot D*: four-loop level ($\ell = 4$, $1 \le j \le 4$), the functions $\varrho_3^{(4)}(\sigma)$ and $\varrho_4^{(4)}(\sigma)$ are scaled by the factors of 10 and 10^2, respectively. The values of parameters: $n_f = 3$ active flavors, $y = \ln(\sigma/\Lambda^2)$.

couplant $a_s^{(\ell)}(Q^2)$, are shown in Fig. 4.9. As one can infer from this figure, the functions $\varrho_j^{(\ell)}(\sigma)$ vanish at both $\sigma \to \infty$ and $\sigma \to 0$. The first-order functions ($j = 1$) assume positive values, whereas the higher-order functions ($j \ge 2$) may also assume negative values. In general, the partial spectral function $\varrho_j^{(\ell)}(\sigma)$ changes its sign ($j - 1$) times on the interval $0 < \sigma < \infty$. It is also worthwhile to mention that the higher-order functions $\varrho_j^{(\ell)}(\sigma)$ ($j \ge 2$) appear to be substantially suppressed with respect to those of the preceding orders. For example, Fig. 4.9D implies that at the four-loop level the maximum value of the fourth-order function $\varrho_4^{(4)}(\sigma)$ is two orders of magnitude

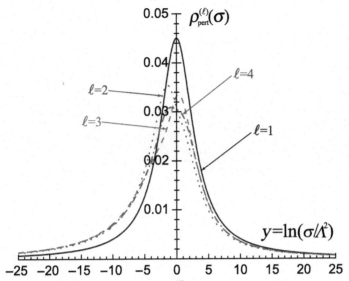

FIG. 4.10 The perturbative spectral function $\rho_{\text{pert}}^{(\ell)}(\sigma)$ (4.58) at various loop levels ($1 \le \ell \le 4$). The values of parameters: $n_f = 3$ active flavors, $y = \ln(\sigma/\Lambda^2)$.

less than the maximum value of the first-order function $\varrho_1^{(4)}(\sigma)$. In turn, the fact that the function $\varrho_{j+1}^{(\ell)}(\sigma)$ is subdominant to $\varrho_j^{(\ell)}(\sigma)$ eventually leads to an enhanced stability[13] of the approach on hand with respect to both higher loop corrections and applied subtraction scheme at low and moderate energies (see Chapter 5 for details).

The plots of the perturbative spectral function $\rho_{\text{pert}}^{(\ell)}(\sigma)$ (4.58) at first four loop levels ($1 \le \ell \le 4$) are shown in Fig. 4.10. As one can infer from this figure, the function (4.58) assumes positive values. Additionally, Figs. 4.9 and 4.10 imply that the function $\rho_{\text{pert}}^{(\ell)}(\sigma)$ vanishes at both $\sigma \to \infty$ and $\sigma \to 0$. In particular, in the ultraviolet asymptotic $\sigma \to \infty$ the one-loop function (4.58) takes the form

$$\rho_{\text{pert}}^{(1)}(\sigma) \simeq \frac{d_1}{y^2} + \mathcal{O}\left(\frac{1}{y^4}\right), \quad y = \ln\left(\frac{\sigma}{\Lambda^2}\right), \quad y \to \infty, \tag{4.102}$$

whereas at the higher loop levels ($\ell \ge 2$)

$$\rho_{\text{pert}}^{(\ell)}(\sigma) \simeq \frac{d_1}{y^2} + \frac{d_1 B_1(1 - 2\ln y) + 2d_2}{y^3} + \mathcal{O}\left(\frac{1}{y^4}\right), \quad y \to \infty. \tag{4.103}$$

The qualitative behavior of the perturbative spectral function $\rho_{\text{pert}}^{(\ell)}(\sigma)$ at $\sigma \to 0$ turns out to be quite similar to that of at $\sigma \to \infty$, except for the fact that at

---

13. With respect to that of the perturbative approach.

the higher loop levels the leading-order term is slightly altered with respect to the one-loop level. Specifically, in the infrared asymptotic $\sigma \to 0$ the one-loop spectral function (4.58) can be represented as

$$\rho_{\text{pert}}^{(1)}(\sigma) \simeq \frac{d_1}{y^2} + \mathcal{O}\left(\frac{1}{y^4}\right), \quad y = \ln\left(\frac{\sigma}{\Lambda^2}\right), \quad y \to -\infty. \tag{4.104}$$

In turn, the two-loop correction affects the leading-order term of the expansion of the spectral function at $\sigma \to 0$

$$\rho_{\text{pert}}^{(2)}(\sigma) \simeq \frac{d_1(1 + B_1)}{y^2} + \frac{d_1 B_1(1 - 2\ln y) + 2d_2(1 + B_1)}{y^3} + \mathcal{O}\left(\frac{1}{y^4}\right),$$

$$y \to -\infty, \tag{4.105}$$

whereas at the higher loop levels ($\ell \geq 3$), the first two terms of the expansion of $\rho_{\text{pert}}^{(\ell)}(\sigma)$ remain the same:

$$\rho_{\text{pert}}^{(\ell)}(\sigma) \simeq \frac{d_1(1 + B_1)}{y^2} + \frac{1 + B_1}{y^3}[d_1 B_1(1 - 2\ln y) + 2d_2] + \mathcal{O}\left(\frac{1}{y^4}\right),$$

$$y \to -\infty. \tag{4.106}$$

Eqs. (4.102)–(4.106) are also illustrated by Fig. 4.10, which, in particular, indicates that the range of y, where the difference between ℓ-loop and $(\ell + 1)$-loop perturbative spectral function (4.58) is sizable, is located in the vicinity of $y = 0$, and becomes smaller at larger ℓ.

It also has to be mentioned that the perturbative spectral function $\rho_{\text{pert}}(\sigma)$ (4.55) provides only a partial information on the dynamics of a physical process on hand. Specifically, the definition (4.55) retains a purely perturbative part of the hadronic vacuum polarization function $\Pi(q^2)$, which is represented by the diagrams of the type of those displayed in Figs. 3.2 and 3.3. In particular, this implies that, for example, the vacuum condensate contributions to $\Pi(q^2)$, which are represented by the diagrams of the type of those displayed in Fig. 3.5, as well as the effects due to other intrinsically nonperturbative mechanisms, which may ultimately govern the confinement of colored fields, are not included into Eq. (4.55). Nonetheless, as it will be demonstrated in Chapter 5, the unified integral representations (4.34)–(4.36) prove to be capable of providing physically sound results for the functions on hand in the energy range substantially extended with respect to that of the QCD perturbation theory, even if such abridged input as the perturbative spectral function (4.55) is employed.

4.3 NOTE ON THE MASSLESS LIMIT

First of all it is necessary to emphasize that the effects due to the nonvanishing mass of the lightest hadronic state can be safely neglected only at high energies, whereas at low energies such effects play a valuable role. For example, in the massless limit the leading-order term of the hadronic vacuum polarization function (3.32) implies that the electron-positron pair is capable of annihilating into hadrons irrespective of what the center-of-mass energy squared s is (i.e., for $s \geq 0$), though this process is kinematically forbidden for $0 \leq s < m^2$. The very same thing can also be said about the τ lepton hadronic decay. Specifically, the nonvanishing value of the respective hadronic production threshold, which exceeds the masses of electron and muon, but is less than the mass of the τ lepton ($m_e < m_\mu < m < m_\tau$), implements the experimental fact that the τ lepton is the only lepton, which is heavy enough ($m_\tau \simeq 1.777\,\text{GeV}$ [84]) to decay into hadrons. However, in the massless limit ($m = 0$) the leading-order term of the theoretical prediction for the hadronic decay width of a lepton turns out to be nonvanishing not only for τ lepton, but also for muon and electron, that is certainly kinematically forbidden, see the discussion of this issue in paper [337] and references therein. Additionally, in the massless limit some of the intrinsically nonperturbative constraints, which relevant dispersion relations impose on the functions on hand, appear to be lost.

Specifically, in the massless limit ($m = 0$) the dispersion relations (4.1)–(4.6) impose the set of nonperturbative constraints on the functions $\Pi(q^2)$, $R(s)$, and $D(Q^2)$, which differs from that of discussed in Section 4.1. In particular, in this case the dispersion relation (4.1) implies that the hadronic vacuum polarization function $\Pi(q^2)$ has the only cut $q^2 \geq 0$ along the positive semiaxis of real q^2 and that the points q^2 and q_0^2 can be located anywhere in the complex q^2-plane except for this cut.[14] In turn, Eqs. (4.4), (4.6) signify that the function $R(s)$ assumes nonzero values for any center-of-mass energy squared $s \geq 0$ (that contradicts the aforementioned kinematic restrictions) and accounts for the effects due to continuation of spacelike theoretical results into the timelike domain. And finally the dispersion relation (4.3) indicates that the Adler function $D(Q^2)$ has the only cut $Q^2 \leq 0$ along the negative semiaxis of real Q^2. As for the nonperturbative constraint $D(Q^2) \to 0$ at $Q^2 \to 0$, which essentially stabilizes the infrared behavior of the Adler function and plays a substantial role in the study of hadron dynamics at low energies, it appears to be lost in the massless limit.

To express the functions $\Pi(q^2)$, $R(s)$, and $D(Q^2)$ in terms of the common spectral density in the massless limit one can either repeat the same steps[15] as those performed in Section 4.1 or simply put $m = 0$ in Eqs. (4.34)–(4.36), which eventually leads to

$$\Delta\Pi(q^2, q_0^2) = -\ln\left(\frac{-q^2}{-q_0^2}\right) + \int_0^\infty \rho(\sigma) \ln\left[\frac{1 - (\sigma/q^2)}{1 - (\sigma/q_0^2)}\right] \frac{d\sigma}{\sigma}, \qquad (4.107)$$

---

14. The implications of this condition in the massless limit, such as the infrared divergence of the hadronic vacuum polarization function $\Pi(q^2)$, will be discussed in Section 5.1.

15. Assuming $m = 0$ and using in Eq. (4.6) the integration contours displayed in Fig. 3.7.

$$R(s) = \theta(s) \left[1 + \int_s^\infty \rho(\sigma) \frac{d\sigma}{\sigma} \right], \qquad (4.108)$$

$$D(Q^2) = 1 + \int_0^\infty \frac{\rho(\sigma)}{\sigma + Q^2} d\sigma, \qquad (4.109)$$

where $\rho(\sigma)$ is determined in Eq. (4.37). Note that the leading-order terms of the functions (4.107)–(4.109) coincide with those of their perturbative approximations (3.31), (3.53), (3.61), which are evidently valid at high energies only. As discussed earlier, the massless integral representations (4.107)–(4.109) retain only a part of the nonperturbative constraints, which dispersion relations (4.1)–(4.6) impose on the functions on hand, while an essential part of those constraints is missing. This fact eventually results in a physically disfavored behavior of the functions (4.107)–(4.109) at low energies, which will be illustrated in the next chapter. In particular, in the case of perturbative spectral function (4.55) the hadronic vacuum polarization function (4.107) diverges in the infrared domain, the R-ratio (4.108) does not vanish for the energies below the hadronic production threshold, and the Adler function (4.109) assumes a finite value in the limit $Q^2 \to 0$. At the same time the representations (4.107)–(4.109) appear to be useful for the analysis of hadron dynamics at moderate and high energies. Specifically, Eqs. (4.107)–(4.109) eliminate unphysical perturbative singularities and somewhat extend the range of applicability of the QCD perturbation theory toward the low energies. The representation (4.108) accounts for the effects due to the continuation of the spacelike theoretical results into the timelike domain. Additionally, Eqs. (4.107)–(4.109) yield the results of an enhanced stability, with respect to that of the QCD perturbation theory, to both the higher loop corrections and the choice of the subtraction scheme at moderate and low energies (see also Chapter 5).

It has to be noted that in the case of perturbative spectral function (4.55) the massless integral representations for the function $R(s)$ (4.108) and for the Adler function $D(Q^2)$ (4.109) become identical to those of the "analytic perturbation theory" (APT) proposed in late 1990s by Shirkov, Solovtsov, and Milton in papers [331, 353–355], whereas the hadronic vacuum polarization function $\Pi(q^2)$ was not addressed in the framework of the latter. The basic idea of APT was to impose the spectral representation of the Källén-Lehmann type on the strong running coupling $\alpha_s(Q^2)$ itself and on all its integer[16] powers appearing in the perturbative approximations of the physical observables depending on the spacelike kinematic variable (which constitutes the so-called "analytization" procedure), to postulate the definition of the spectral function in

---

16. A generalization of the analytic perturbation theory to noninteger powers of the strong running coupling was presented in papers [356–359].

the form of Eq. (4.55c), and to use the relation (4.6) to continue the resulting expression into the timelike domain. The APT and its modifications have been applied to the study of such issues of elementary particle physics as deep inelastic scattering and the corresponding sum rules (papers [360–368] and [369–373], respectively), pion electromagnetic form factor [374–377], QCD bound states [345, 378–385], inclusive τ lepton hadronic decay[17] [346, 391–393], chiral symmetry breaking [335, 394, 395], quark-gluon plasma [396–398], and some others (see, for example, papers [399–404]). It is also worthwhile to mention that, as argued in papers [347, 348, 405–407], the APT can be somewhat ameliorated, but only at the expense of engaging additional phenomenological assumptions. The detailed overview of the analytic perturbation theory and its applications can be found in papers [195, 408–411] and references therein.

It is necessary to outline that a rigorous unabridged theoretical framework, which would have enabled one to thoroughly account for all the tangled nonperturbative dynamics of colored fields that ultimately results in the formation of bound states of quarks, as well as to accurately assess the masses of such bound states, is still far from being feasible. However, this mere fact by no means can serve as a justification neither for neglecting the effects due to the nonvanishing mass m of the lightest hadronic state, nor for assigning to the latter any value other than the experimentally measured one. Basically, the effects due to the nonvanishing m play an essential role in the study of hadron dynamics at low energies and cannot be disregarded, unless the energy of a process on hand is high enough to securely make them negligible. In particular, as it will be discussed in Chapter 5 the proper account for such effects eventually leads to a physically sound behavior of the functions on hand in the infrared domain, that, in turn, is decisive for the theoretical description of a variety of the strong interaction processes at low energies.

---

17. The APT and its modifications basically examine either the total sum of vector and axial-vector terms of the inclusive semileptonic branching ratio or the vector term only. As discussed in papers [337, 343, 344, 386–390], for a self-consistent description of inclusive τ lepton hadronic decay in vector and axial-vector channels the effects due to nonvanishing hadronic production threshold have to be properly accounted for.

Chapter 5

Functions $\Pi(q^2)$, $R(s)$, and $D(Q^2)$ Within Dispersive Approach

The dispersive approach to QCD provides the unified integral representations for the hadronic vacuum polarization function $\Pi(q^2)$, the R-ratio of electron-positron annihilation into hadrons $R(s)$, and the Adler function $D(Q^2)$. These representations merge, in a self-consistent way, the intrinsically nonperturbative constraints, which originate in the pertinent kinematic restrictions, with corresponding perturbative input. The unified integral representations for the functions $\Pi(q^2)$, $R(s)$, and $D(Q^2)$ contain no unphysical singularities and properly account for the effects due to continuation of the spacelike theoretical results into the timelike domain, which appear to dominate over corresponding perturbative contributions. At the same time the unified integral representations for the functions on hand coincide with respective perturbative approximations at high energies and possess an enhanced stability, with respect to that of the perturbative approach, to the higher loop corrections and to the choice of the subtraction scheme at low and moderate energies. All these features make the dispersively improved perturbation theory (DPT) capable of handling the strong interaction processes in the energy range substantially extended with respect to that of the perturbative approach.

5.1 HADRONIC VACUUM POLARIZATION FUNCTION WITHIN DPT

As it was delineated in the previous chapter the dispersive approach to QCD provides the following integral representation for the hadronic vacuum polarization function $\Pi(q^2)$ (4.34), which contains no unphysical singularities and properly accounts for all the relevant nonperturbative constraints discussed in Section 4.1:

$$\Delta\Pi(q^2, q_0^2) = \Delta\Pi^{(0)}(q^2, q_0^2) + \int_{m^2}^{\infty} \rho(\sigma)\ln\left(\frac{\sigma - q^2}{\sigma - q_0^2}\frac{m^2 - q_0^2}{m^2 - q^2}\right)\frac{d\sigma}{\sigma}. \quad (5.1)$$

Strong Interactions in Spacelike and Timelike Domains. http://dx.doi.org/10.1016/B978-0-12-803439-2.00005-3

In this equation $\Delta\Pi(q^2, q_0^2) = \Pi(q^2) - \Pi(q_0^2)$ is the subtracted hadronic vacuum polarization function, q_0^2 stands for the subtraction point,[1] m denotes the total mass of the respective lightest allowed hadronic final state, the leading-order term reads (4.39)

$$\Delta\Pi^{(0)}(q^2, q_0^2) = 2\frac{\varphi - \tan\varphi}{\tan^3\varphi} - 2\frac{\varphi_0 - \tan\varphi_0}{\tan^3\varphi_0}, \tag{5.2}$$

$\sin^2\varphi = q^2/m^2$, $\sin^2\varphi_0 = q_0^2/m^2$, and $\rho(\sigma)$ is the spectral density (4.37), which provides the corresponding perturbative input. As outlined in Section 4.2 the perturbative part of the spectral density (4.55)

$$\rho_{\text{pert}}^{(\ell)}(\sigma) = \sum_{j=1}^{\ell} d_j \varrho_j^{(\ell)}(\sigma) \tag{5.3}$$

can be explicitly calculated at an arbitrary loop level, assuming that the involved coefficients of the perturbative expansion of the Adler function and of the renormalization group β function are available. In Eq. (5.3) $\varrho_j^{(\ell)}(\sigma)$ denotes the partial spectral function (4.65), which corresponds to the jth power of the ℓ-loop perturbative couplant:

$$\varrho_j^{(\ell)}(\sigma) = \frac{1}{2\pi i} \lim_{\varepsilon \to 0_+} \left\{ \left[a_s^{(\ell)}(-\sigma - i\varepsilon) \right]^j - \left[a_s^{(\ell)}(-\sigma + i\varepsilon) \right]^j \right\}. \tag{5.4}$$

Basically, at any given loop level the integration on the right-hand side of Eq. (5.1) can be performed for the whole spectral function (5.3) in the way described in Appendix B. At the same time for illustrative purposes it is worthwhile to handle each term of the sum on the right-hand side of Eq. (5.3) separately from the other terms. In this case the ℓ-loop strong correction to the hadronic vacuum polarization function (5.1) can be represented as

$$\Delta p^{(\ell)}(q^2, q_0^2) = \sum_{j=1}^{\ell} d_j \Delta A_{\text{HP},j}^{(\ell)}(q^2, q_0^2) \tag{5.5}$$

with

$$\Delta A_{\text{HP},j}^{(\ell)}(q^2, q_0^2) = \int_{m^2}^{\infty} \varrho_j^{(\ell)}(\sigma) \ln\left(\frac{\sigma - q^2}{\sigma - q_0^2} \frac{m^2 - q_0^2}{m^2 - q^2} \right) \frac{d\sigma}{\sigma} \tag{5.6}$$

being an effective ℓ-loop expansion function of the jth order.

Thus at the ℓ-loop level the DPT yields the following expression for the subtracted hadronic vacuum polarization function:

---

1. In Eq. (5.1) the points q^2 and q_0^2 can be located anywhere in the complex q^2-plane except for the cut $q^2 \geq m^2$ along the positive semiaxis of real q^2.

$$\Delta\Pi^{(\ell)}(q^2, q_0^2) = \Delta\Pi^{(0)}(q^2, q_0^2) + \sum_{j=1}^{\ell} d_j \Delta A_{HP,j}^{(\ell)}(q^2, q_0^2), \tag{5.7}$$

where the leading-order term and the higher-order expansion functions are given in Eqs. (5.2), (5.6), respectively. It is worth noting that in the massless limit ($m = 0$) the DPT expression for $\Pi(q^2)$ (5.7) acquires the form

$$\Delta\Pi_{APT}^{(\ell)}(q^2, q_0^2) = -\ln\left(\frac{-q^2}{-q_0^2}\right) + \sum_{j=1}^{\ell} d_j \int_0^\infty \varrho_j^{(\ell)}(\sigma) \ln\left(\frac{-q_0^2}{-q^2}\frac{\sigma - q^2}{\sigma - q_0^2}\right) \frac{d\sigma}{\sigma}, \tag{5.8}$$

which formally corresponds to the analytic perturbation theory, though the hadronic vacuum polarization function was not addressed in the framework of the latter, see discussion of this issue in Section 4.3. At the same time the perturbative approximation of $\Pi(q^2)$ reads

$$\Delta\Pi_{pert}^{(\ell)}(q^2, q_0^2) = -\ln\left(\frac{-q^2}{-q_0^2}\right) + \Delta p_{pert}^{(\ell)}(q^2, q_0^2), \tag{5.9}$$

see Eq. (3.34) and Section 3.1. Evidently, the APT expression (5.8) and the perturbative approximation (5.9) diverge in the infrared limit, which makes them both inapplicable at low energies.

In what follows it proves to be convenient to deal with the subtracted at zero form of the hadronic vacuum polarization function. The DPT expression (5.1) for $q_0^2 = 0$ takes the following form:

$$\bar{\Pi}(Q^2) \equiv \Delta\Pi(0, -Q^2) = \Delta\Pi^{(0)}(0, -Q^2) + \int_{m^2}^\infty \rho(\sigma) \ln\left(\frac{1 + Q^2/m^2}{1 + Q^2/\sigma}\right) \frac{d\sigma}{\sigma}, \tag{5.10}$$

where $Q^2 = -q^2 \geq 0$ stands for the spacelike kinematic variable. As mentioned above at the ℓ-loop level the second term on the right-hand side of this equation can also be represented as

$$\bar{p}^{(\ell)}(Q^2) \equiv \Delta p^{(\ell)}(0, -Q^2) = \sum_{j=1}^{\ell} d_j A_{HP,j}^{(\ell)}(Q^2), \tag{5.11}$$

where

$$A_{HP,j}^{(\ell)}(Q^2) \equiv \Delta A_{HP,j}^{(\ell)}(0, -Q^2) = \int_{m^2}^\infty \varrho_j^{(\ell)}(\sigma) \ln\left(\frac{1 + Q^2/m^2}{1 + Q^2/\sigma}\right) \frac{d\sigma}{\sigma} \tag{5.12}$$

is the corresponding ℓ-loop effective expansion function of the jth order and $\varrho_j^{(\ell)}(\sigma)$ stands for the partial spectral function (5.4). It is necessary to outline that the APT expression for the hadronic vacuum polarization function (5.8) and the perturbative approximation (5.9) possess the analytic properties in the kinematic variable q^2, which differ from those of prescribed by the dispersion

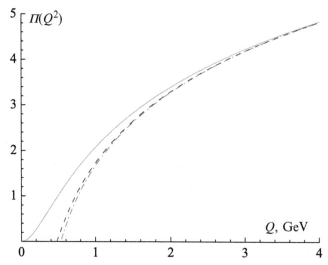

FIG. 5.1 The one-loop hadronic vacuum polarization function $\Pi(Q^2)$: DPT expression (Eq. 5.10, *solid curve*), its massless limit (Eq. 5.16, *dashed curve*), and the perturbative approximation (Eq. 5.17, *dot-dashed curve*). The values of parameters: $n_f = 3$ active flavors, $\Lambda = 350$ MeV.

relation (4.1). As it will be discussed later one of the implications of this fact is that neither APT expression (5.8) nor perturbative approximation (5.9) can be subtracted at the point $q_0^2 = 0$ because they both diverge at low energies.

Let us begin by addressing the hadronic vacuum polarization function at the one-loop level ($\ell = 1$). In this case the pertinent strong correction (5.11) consists of only one term, specifically,

$$\bar{p}^{(1)}(Q^2) = d_1 A_{HP,1}^{(1)}(Q^2), \tag{5.13}$$

where $d_1 = 4/\beta_0$ and $\beta_0 = 11 - 2n_f/3$ denote the one-loop coefficients of perturbative expansions of the Adler function (see Section 3.2) and β function (see Section 2.1), respectively, n_f stands for the number of active flavors, and

$$A_{HP,1}^{(1)}(Q^2) = \int_{m^2}^{\infty} \varrho_1^{(1)}(\sigma) \ln\left(\frac{1 + Q^2/m^2}{1 + Q^2/\sigma}\right) \frac{d\sigma}{\sigma}. \tag{5.14}$$

In this equation $\varrho_1^{(1)}(\sigma)$ is the one-loop spectral function (4.63)

$$\varrho_1^{(1)}(\sigma) = \frac{1}{\ln^2(\sigma/\Lambda^2) + \pi^2} \tag{5.15}$$

and Λ is the QCD scale parameter. The integration in Eq. (5.14) can be performed numerically in the way described in Appendix B.

The plot of the one-loop hadronic vacuum polarization function (5.10) is presented in Fig. 5.1 (solid curve). To elucidate the qualitative distinctions between various approaches mentioned earlier, Fig. 5.1 also displays the one-loop massless expression (5.8), which, in the considered case, corresponds to the result of the analytic perturbation theory discussed in Section 4.3 (dashed curve)

$$\Delta\Pi_{\mathrm{APT}}^{(1)}(-Q_0^2,-Q^2) = \ln\left(\frac{Q^2}{Q_0^2}\right) + \frac{4}{\beta_0}\ln\left[\frac{a_{\mathrm{an}}^{(1)}(Q_0^2)}{a_{\mathrm{an}}^{(1)}(Q^2)}\right], \qquad (5.16)$$

and the one-loop perturbative approximation of $\Pi(q^2)$ (dot-dashed curve)

$$\Delta\Pi_{\mathrm{pert}}^{(1)}(-Q_0^2,-Q^2) = \ln\left(\frac{Q^2}{Q_0^2}\right) + \frac{4}{\beta_0}\ln\left[\frac{a_{\mathrm{s}}^{(1)}(Q_0^2)}{a_{\mathrm{s}}^{(1)}(Q^2)}\right]. \qquad (5.17)$$

As earlier, in Eqs. (5.16), (5.17) the notation for the couplant $a(Q^2) = \alpha(Q^2)\beta_0/(4\pi)$ is used,

$$\alpha_{\mathrm{s}}^{(1)}(Q^2) = \frac{4\pi}{\beta_0}\frac{1}{\ln z}, \qquad z = \frac{Q^2}{\Lambda^2} \qquad (5.18)$$

denotes the one-loop perturbative running coupling (2.19), and

$$\alpha_{\mathrm{an}}^{(1)}(Q^2) = \frac{4\pi}{\beta_0}\frac{z-1}{z\ln z} \qquad (5.19)$$

stands for the one-loop infrared enhanced analytic running coupling [345, 346]. The latter contains no unphysical singularities in the kinematic variable Q^2, and, as one can see, diverges in the infrared limit $Q^2 \to 0$. It is interesting to note here that the expression (5.19) was first obtained in papers [345, 346], independently rediscovered (proceeding from an entirely different reasoning) later on in Refs. [412, 413], and rederived in a paper [379] afterwards. Specifically, in papers [412, 413] it has been revealed that the running coupling $\alpha_{\mathrm{an}}^{(1)}(Q^2)$ (5.19) obeys a remarkable symmetry relation, which binds its behavior at high and low energies. Namely, the invariant charge (5.19), being multiplied by $\sqrt{z}$, appears to be invariant under transformation $z \to 1/z$:

$$\sqrt{\frac{Q^2}{\Lambda^2}}\alpha_{\mathrm{an}}^{(1)}(Q^2) = \sqrt{\frac{\Lambda^2}{Q^2}}\alpha_{\mathrm{an}}^{(1)}\left(\frac{\Lambda^4}{Q^2}\right). \qquad (5.20)$$

It is this feature of the strong running coupling (5.19) that explicitly reproduces the conformal inversion symmetry of the instanton size distribution observed in the corresponding results of the UKQCD Collaboration [414] (see papers [195, 412, 413, 415] and references therein for the details). The properties of the invariant charge (5.19) have been studied in Refs. [195, 345, 346], whereas some of its recent applications can be found in, for example, papers [396–398].

As one can infer from Fig. 5.1, the perturbative approximation of the hadronic vacuum polarization function (5.17) diverges at a positive Q^2, which invalidates it at low energies. At the same time although both DPT and APT expressions, Eqs. (5.10), (5.8), respectively, contain no unphysical singularities

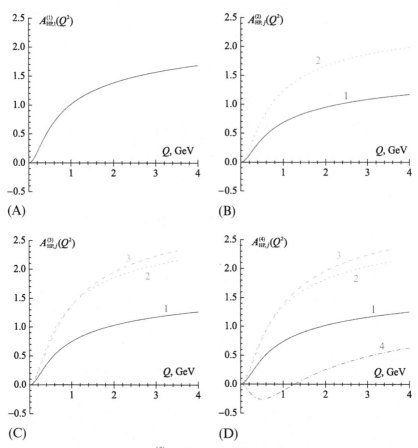

FIG. 5.2   Expansion functions $A_{\text{HP},j}^{(\ell)}(Q^2)$ (5.12) at various loop levels ($1 \le \ell \le 4$, $1 \le j \le \ell$). *Plot A*: one-loop level ($\ell = 1$, $j = 1$). *Plot B*: two-loop level ($\ell = 2$, $1 \le j \le 2$), the function $A_{\text{HP},2}^{(2)}(Q^2)$ is scaled by the factor of 10. *Plot C*: three-loop level ($\ell = 3$, $1 \le j \le 3$), the functions $A_{\text{HP},2}^{(3)}(Q^2)$ and $A_{\text{HP},3}^{(3)}(Q^2)$ are scaled by the factors of 10 and 10^2, respectively. *Plot D*: four-loop level ($\ell = 4$, $1 \le j \le 4$), the functions $A_{\text{HP},2}^{(4)}(Q^2)$, $A_{\text{HP},3}^{(4)}(Q^2)$, and $A_{\text{HP},4}^{(4)}(Q^2)$ are scaled by the factors of 10, 10^2, and 10^3, respectively. *Numerical labels* specify the order j of the ℓ-loop function $A_{\text{HP},j}^{(\ell)}(Q^2)$. The values of parameters: $n_f = 3$ active flavors, $\Lambda = 350\,\text{MeV}$.

at positive values of real Q^2, their qualitative behavior in the infrared domain appears to be essentially different. Specifically, the DPT expression for the hadronic vacuum polarization function (5.10) provides physically sound behavior of $\Pi(q^2)$ in the entire energy range, whereas the APT prediction (5.8) diverges in the infrared limit, that makes it inapplicable at low energies. Note that the divergence of the APT prediction for $\Pi(q^2)$ originates in the mathematical fact that in the massless limit the hadronic vacuum polarization

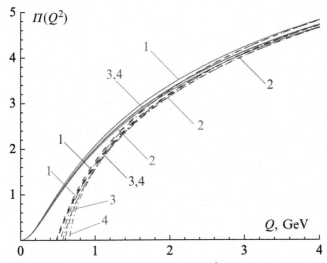

FIG. 5.3 The hadronic vacuum polarization function $\Pi(Q^2)$ at various loop levels ($1 \leq \ell \leq 4$): DPT expression (Eq. 5.7, *solid curves*), its massless limit (Eq. 5.8, *dashed curves*), and the perturbative approximation (Eq. 5.9, *dot-dashed curves*). *Numerical labels* specify the loop level. The values of parameters: $n_f = 3$ active flavors, $\Lambda = 350\,\text{MeV}$.

function $\Pi(q^2)$ possesses the only cut $q^2 \geq 0$ along the positive semiaxis of real q^2, and appears to be undefined at the beginning of its branch cut irrespective of which admissible value is assigned to the subtraction point q_0^2. It is worthwhile to mention here that the aforementioned distinctive features of the function $\Pi(q^2)$ are universal and specify its qualitative behavior within each of the approaches discussed earlier.

Let us proceed now to the hadronic vacuum polarization function at the higher loop levels ($\ell \geq 2$). As mentioned earlier in this case the corresponding strong correction (5.11) can be represented as the sum of the ℓ-loop expansion functions (5.12), specifically,

$$\bar{p}^{(\ell)}(Q^2) = d_1 A_{\text{HP},1}^{(\ell)}(Q^2) + d_2 A_{\text{HP},2}^{(\ell)}(Q^2) + d_3 A_{\text{HP},3}^{(\ell)}(Q^2) + \cdots. \tag{5.21}$$

The number of terms on the right-hand side of this equation is determined by the loop level. Similar to the one-loop level the expansion functions $A_{\text{HP},j}^{(\ell)}(Q^2)$ can be calculated numerically since the involved partial spectral functions $\varrho_j^{(\ell)}(\sigma)$ (5.4) turn out to be rather cumbersome (see Appendix B).

The plots of the expansion functions $A_{\text{HP},j}^{(\ell)}(Q^2)$ (5.12) up to the four-loop level ($1 \leq \ell \leq 4$, $1 \leq j \leq \ell$) are displayed in Fig. 5.2. As one can infer from this figure the expansion functions $A_{\text{HP},j}^{(\ell)}(Q^2)$ (5.12) contain no unphysical singularities, vanish in the infrared limit $Q^2 \to 0$, and tend to the respective perturbative expressions at high energies. The expansion functions of the first

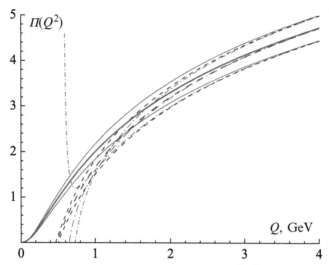

FIG. 5.4 The four-loop hadronic vacuum polarization function $\Pi(Q^2)$ in the "MS-like" subtraction scheme: DPT expression (Eq. 5.7, *solid curves*), its massless limit (Eq. 5.8, *dashed curves*), and the perturbative approximation (Eq. 5.9, *dot-dashed curves*). *Thick curves* denote the results corresponding to the $\overline{\text{MS}}$-scheme, whereas *thin curves* portray the alteration of $\Pi(Q^2)$ due to variability of the involved scheme-dependent coefficients. The values of parameters: $n_f = 3$ active flavors, $\Lambda = 350\,\text{MeV}$.

orders ($1 \leq j \leq 3$) assume positive values, whereas the higher-order functions ($j \geq 4$) may also assume negative values. Fig. 5.2 implies that the expansion functions $A_{\text{HP},j}^{(\ell)}(Q^2)$ of the higher orders ($j \geq 2$) are subdominant to the functions of the preceding orders. In particular, Fig. 5.2D indicates that the expansion function of the $(j+1)$th order $A_{\text{HP},j+1}^{(\ell)}(Q^2)$ is suppressed by an order of magnitude with respect to the function of the jth order $A_{\text{HP},j}^{(\ell)}(Q^2)$ in the entire energy range. In turn, this fact eventually results in an enhanced stability of the hadronic vacuum polarization function obtained within the dispersive approach, with respect to that of the QCD perturbation theory, to the higher loop corrections and the choice of the subtraction scheme at intermediate and low energies.

In particular, the higher loop stability of the dispersively improved perturbation theory is illustrated by Fig. 5.3, which displays the hadronic vacuum polarization function $\Pi(q^2)$ calculated within various approaches at first four loop levels. As one can infer from this figure, at low energies the variability of the DPT expression for $\Pi(q^2)$ (Eq. 5.7, solid curves) caused by the incorporation of the higher loop corrections is considerably less than that of the APT prediction (Eq. 5.8, dashed curves) and substantially less than that of the perturbative approximation (Eq. 5.9, dot-dashed curves). This effect takes place due to the fact that, as mentioned above, the APT prediction for the hadronic vacuum polarization function $\Pi(q^2)$ diverges at $q^2 \to 0$, while its perturbative approximation contains unphysical singularities at negative values of real q^2, that makes the higher loop contributions to $\Pi(q^2)$ enhanced at low energies. At

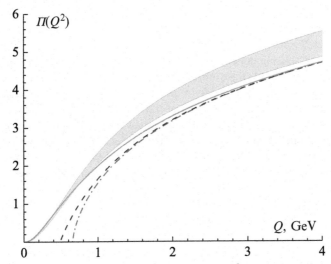

FIG. 5.5 The hadronic vacuum polarization function $\Pi(Q^2)$: the experimental prediction (Eq. 1.43, *shaded band*), the DPT expression (Eq. 5.7, *solid curve*), its massless limit (Eq. 5.8, *dashed curve*), and the perturbative approximation (Eq. 5.9, *dot-dashed curve*). The presented results correspond to the four-loop level. The values of parameters: $n_f = 3$ active flavors, $\Lambda = 350\,\mathrm{MeV}$.

the same time, at intermediate energies the distinctions between the approaches on hand become less pronounced, whereas in the ultraviolet asymptotic all three approaches yield the same result.

The enhanced stability of the dispersive approach to the choice of the subtraction scheme is illustrated by Fig. 5.4. In particular, this figure displays the hadronic vacuum polarization function $\Pi(q^2)$ calculated within approaches on hand at the four-loop level and elucidates its sensitivity to the variation of the involved scheme-dependent coefficients of the perturbative expansions of the Adler function (d_2, d_3, and d_4) and renormalization group β function (β_2 and β_3) (see Sections 3.2 and 2.1, respectively). Specifically, in Fig. 5.4 the DPT expression for the hadronic vacuum polarization function (5.7) is shown by solid curves, the APT prediction (5.8) is denoted by dashed curves, and the perturbative approximation (5.9) is marked by dot-dashed curves. For either of the approaches the internal (thick) curve corresponds to the $\overline{\mathrm{MS}}$-scheme, whereas the external (thin) curves delimitate the variability range of $\Pi(q^2)$ caused by a probe alteration of the involved scheme-dependent coefficients within "MS-like" scheme. Fig. 5.4 implies that at low and moderate energies the DPT expression for the hadronic vacuum polarization function (5.7) possesses a somewhat higher stability with respect to the variation of the corresponding scheme-dependent coefficients than that of its massless limit (5.8). As for the perturbative approximation of $\Pi(q^2)$, its scheme stability is substantially less than those of the DPT and APT expressions. Again such effect takes place due to the divergence of the hadronic vacuum polarization function calculated within perturbative approach (at a real negative value of q^2) and within APT (in the infrared limit $q^2 \to 0$) at any given loop level, that, in turn, amplifies the corresponding scheme-dependent contributions at low energies. As earlier all three approaches provide the expressions for $\Pi(q^2)$, which tend to each other at intermediate energies and coincide in the ultraviolet asymptotic.

Fig. 5.5 presents the juxtaposition of the four-loop hadronic vacuum polarization function $\Pi(q^2)$ calculated in the framework of the approaches on hand with its experimental prediction delineated in Section 1.2. As one can infer from this figure, the perturbative approximation of the function $\Pi(q^2)$ (Eq. 5.9, dot-dashed curve) fails to describe the hadronic vacuum polarization function at low energies. As mentioned earlier this fact is primarily caused by such intrinsic obstacles of the QCD perturbation theory as the infrared unphysical singularities, which invalidate it at low energies. As for the APT prediction for $\Pi(q^2)$ (Eq. 5.8, dashed curve), though it contains no unphysical singularities at negative values of real q^2, it diverges in the infrared limit $q^2 \to 0$, which makes it inapplicable at low energies, too. As discussed earlier this is basically caused by the fact that APT ignores the effects due to the nonvanishing hadronic production threshold, which play valuable role at low energies. At the same time as one can infer from Fig. 5.5, the dispersive approach to QCD provides physically congruous description of $\Pi(q^2)$ (Eq. 5.7, solid curve) in the entire energy range and thereby enables one to handle a variety of the strong interaction processes beyond the limitations of perturbation theory.

5.2 R-RATIO WITHIN DPT

For the R-ratio of electron-positron annihilation into hadrons the dispersive approach to QCD provides the following integral representation (4.35):

$$R(s) = R^{(0)}(s) + \theta(s - m^2) \int\limits_{s}^{\infty} \rho(\sigma) \frac{d\sigma}{\sigma}, \qquad (5.22)$$

which, as discussed earlier, properly accounts for the respective nonperturbative constraints, contains no unphysical singularities, and embodies the effects due to continuation of spacelike theoretical results into the timelike domain. In Eq. (5.22) the leading-order term reads (4.40)

$$R^{(0)}(s) = \theta(s - m^2) \left(1 - \frac{m^2}{s} \right)^{3/2}, \qquad (5.23)$$

$s = q^2 \geq 0$ is the timelike kinematic variable, m denotes the total mass of the lightest allowed hadronic final state, $\theta(x)$ stands for the Heaviside unit step function (1.27), and $\rho(\sigma)$ is the spectral density (4.37). The perturbative part of the latter (4.55), which is explicitly calculable at any given loop level, can be represented as the sum of partial spectral functions $\varrho_j^{(\ell)}(\sigma)$ (5.4), corresponding to the jth power of the ℓ-loop perturbative couplant, namely,

$$\rho_{\text{pert}}^{(\ell)}(\sigma) = \sum_{j=1}^{\ell} d_j \varrho_j^{(\ell)}(\sigma), \qquad (5.24)$$

where d_j denotes the coefficient of perturbative expansion of the Adler function (see Section 3.2). The explicit expressions for the spectral functions $\rho_{\text{pert}}^{(\ell)}(\sigma)$ (5.24) are given in Section 4.2 and Appendix B. As mentioned in the previous section although the integration on the right-hand side of Eq. (5.22) can be performed for the whole spectral function $\rho_{\text{pert}}^{(\ell)}(\sigma)$ in the way delineated in Appendix B, for illustrative purposes it is worthwhile to handle each term of the sum (5.24) separately from the others. In particular, this would allow one to distinguish term-by-term the individual perturbative contributions to Eq. (5.22). Specifically, in this case the ℓ-loop strong correction to the function $R(s)$ (5.22) takes the form

$$r^{(\ell)}(s) = \sum_{j=1}^{\ell} d_j A_{\text{TL},j}^{(\ell)}(s), \qquad (5.25)$$

where

$$A_{\text{TL},j}^{(\ell)}(s) = \theta(s - m^2) \int_s^{\infty} \varrho_j^{(\ell)}(\sigma) \frac{d\sigma}{\sigma} \qquad (5.26)$$

stands for the ℓ-loop effective timelike expansion function of jth order, the latter being the continuation (4.6) of the jth power of ℓ-loop perturbative couplant $[a_s^{(\ell)}(Q^2)]^j$ into the timelike domain $s = q^2 \geq 0$. Obviously, the functions $A_{\text{TL},j}^{(\ell)}(s)$ (5.26) differ from the effective expansion functions $A_{\text{HP},j}^{(\ell)}(Q^2)$ (5.12) discussed in the previous section.

At first glance, Eq. (5.25) resembles the perturbative expression for the strong correction to the Adler function (3.57). However, as it will be revealed later, the function $A_{\text{TL},j}^{(\ell)}(s)$ of the jth order ($j \geq 2$) differs from the jth power of the first-order function $[A_{\text{TL},1}^{(\ell)}(s)]^j$. Therefore Eq. (5.25) constitutes an expansion over the set of functions (5.26) rather than a power series. Moreover, even in the ultraviolet asymptotic $s \to \infty$ the reexpansion of Eq. (5.25) contains, in addition to the terms corresponding to the naive continuation[2] of the perturbative approximation (3.57) into the timelike domain (3.64), an infinite number of the so-called π^2-terms. As it will be discussed in Chapter 6 the latter appear to

---

2. Note that contrary to the DPT expression (5.22) and its massless limit (5.28), both the naive continuation of the spacelike expression (3.57) into the timelike domain (3.64) and the perturbative approximation (5.29) contain infrared unphysical singularities.

be nonnegligible starting from the three-loop level and play a valuable role in the study of R-ratio of electron-positron annihilation into hadrons and related quantities in the entire energy range.

Thus at the ℓ-loop level the dispersively improved perturbation theory yields the following expression for the function $R(s)$:

$$R^{(\ell)}(s) = R^{(0)}(s) + \sum_{j=1}^{\ell} d_j A_{\mathrm{TL},j}^{(\ell)}(s), \qquad (5.27)$$

where the leading-order term is given in Eq. (5.23) and the timelike effective expansion functions $A_{\mathrm{TL},j}^{(\ell)}(s)$ are defined in Eq. (5.26). It is also worthwhile to mention that in the massless limit ($m = 0$) the DPT expression for the R-ratio (5.27) becomes identical to that of the analytic perturbation theory discussed in Section 4.3 (see also papers [416–420]), namely,

$$R_{\mathrm{APT}}^{(\ell)}(s) = 1 + \sum_{j=1}^{\ell} d_j \int_s^{\infty} \varrho_j^{(\ell)}(\sigma) \frac{d\sigma}{\sigma}. \qquad (5.28)$$

It has to be noted that the timelike expansion functions $A_{\mathrm{TL},j}^{(\ell)}(s)$ (5.26) for $s \geq m^2$ coincide with those of the APT approach. At the same time as it will be discussed in the next section, the perturbative approximation of the function $R(s)$, which constitutes the reexpansion of Eq. (5.27) at high energies, takes the following form (3.67)

$$R_{\mathrm{pert}}^{(\ell)}(s) = 1 + \sum_{j=1}^{\ell} r_j \left[a_{\mathrm{s}}^{(\ell)}(|s|) \right]^j, \quad r_j = d_j - \delta_j, \quad s \to \infty. \qquad (5.29)$$

In this equation $a_{\mathrm{s}}^{(\ell)}(Q^2)$ stands for the ℓ-loop perturbative couplant (see Chapter 2 and Appendix A), whereas the coefficients r_j, which account for the π^2-terms, substantially differ from the coefficients of perturbative expansion of the Adler function d_j for $j \geq 3$ (see Chapter 6 for the details).

Let us first address the R-ratio of electron-positron annihilation into hadrons at the one-loop level ($\ell = 1$). In this case the corresponding strong correction (5.25) reads

$$r^{(1)}(s) = d_1 A_{\mathrm{TL},1}^{(1)}(s). \qquad (5.30)$$

In this equation $d_1 = 4/\beta_0$ and $\beta_0 = 11 - 2n_f/3$ stand for the one-loop coefficients of perturbative expansion of the Adler function (see Section 3.2) and β function (see Section 2.1), respectively, n_f denotes the number of active flavors,

$$A_{\mathrm{TL},1}^{(1)}(s) = \theta(s - m^2) \int_s^{\infty} \frac{1}{\ln^2(\sigma/\Lambda^2) + \pi^2} \frac{d\sigma}{\sigma} \qquad (5.31)$$

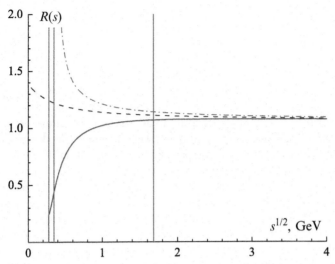

FIG. 5.6 The one-loop function $R(s)$: DPT expression (Eq. 5.27, *solid curve*), its massless limit (Eq. 5.28, *dashed curve*), and the perturbative approximation (Eq. 5.29, *dot-dashed curve*). Vertical *solid lines* mark the locations of (from left to right) the kinematic threshold at $\sqrt{s} = m$, the one-loop perturbative pole at $\sqrt{s} = \Lambda$, and the boundary of the convergence range of the perturbative approximation (5.29) at $\sqrt{s} = \exp(\pi/2)\Lambda$. The values of parameters: $n_f = 3$ active flavors, $\Lambda = 350\,\mathrm{MeV}$.

is the one-loop effective timelike expansion function of the first order (5.26), and Λ stands for the QCD scale parameter. The integration on the right-hand side of Eq. (5.31) can be performed explicitly. Specifically, since (in what follows it is assumed that $\arctan(x)$ is a monotone nondecreasing function of its argument: $-\pi/2 \le \arctan(x) \le \pi/2$ for $-\infty < x < \infty$)

$$\int \frac{1}{\ln^2 x + \pi^2} \frac{dx}{x} = \frac{1}{\pi} \arctan\left(\frac{\ln x}{\pi}\right), \qquad (5.32)$$

one can cast Eq. (5.31) to

$$A_{\mathrm{TL},1}^{(1)}(s) = \theta\left(s - m^2\right)\left[\frac{1}{2} - \frac{1}{\pi}\arctan\left(\frac{\ln w}{\pi}\right)\right], \qquad w = \frac{s}{\Lambda^2}. \qquad (5.33)$$

Basically, this function plays a role[3] of the one-loop effective couplant $a_{\mathrm{TL}}^{(1)}(s)$ in the timelike domain (see also the discussion of the definition of the strong running coupling at low energies given in Section 5.3). Obviously, the one-loop effective "timelike" strong running coupling

$$\alpha_{\mathrm{TL}}^{(1)}(s) = \frac{4\pi}{\beta_0}\theta\left(s - m^2\right)\left\{\frac{1}{2} - \frac{1}{\pi}\arctan\left[\frac{\ln(s/\Lambda^2)}{\pi}\right]\right\} \qquad (5.34)$$

---

3. In general, at any given loop level ℓ the first-order ($j = 1$) expansion function $A_{\mathrm{TL},1}^{(\ell)}(s)$ (5.26) constitutes an effective couplant $a_{\mathrm{TL}}^{(\ell)}(s)$ in the timelike domain.

differs from the naive continuation of the one-loop strong running coupling into the timelike domain (3.64)

$$\alpha_{\text{TL, naive}}^{(1)}(s) \equiv \alpha_s^{(1)}(|s|) = \frac{4\pi}{\beta_0} \frac{1}{\ln(s/\Lambda^2)}, \quad s = q^2 = -Q^2 \geq 0. \quad (5.35)$$

It is interesting to mention here that Eq. (5.34) in the massless limit ($m = 0$) first appeared in early 1980s in paper [330] and only afterwards was derived in Refs. [105, 139, 331]. The detailed discussion of the distinctions between the function $R(s)$ (5.27) and its naive form (3.64) in the ultraviolet domain $s \to \infty$ will be given in Chapter 6 (see also papers [107, 108, 331, 355]).

The one-loop function $R(s)$ calculated within various approaches is presented in Fig. 5.6. As one can infer from this figure the perturbative approximation[4] of the R-ratio (Eq. 5.29, dot-dashed curve) possesses the infrared unphysical singularities, that invalidates it at low energies. As for the DPT expression for $R(s)$ (Eq. 5.27, solid curve) and its massless limit (Eq. 5.28, dashed curve), neither of them contain unphysical singularities. However, the APT prediction (5.28) yields a nonvanishing value of R-ratio for the energies below the hadronic production threshold ($0 \leq s < m^2$), that, as discussed earlier, is kinematically forbidden.

Let us proceed to the function $R(s)$ at the higher loop levels ($\ell \geq 2$). In this case the respective strong correction constitutes the sum (5.25)

$$r^{(\ell)}(s) = d_1 A_{\text{TL},1}^{(\ell)}(s) + d_2 A_{\text{TL},2}^{(\ell)}(s) + d_3 A_{\text{TL},3}^{(\ell)}(s) + \cdots, \quad (5.36)$$

with the number of terms being determined by the loop level. The ℓ-loop expansion function of the jth order $A_{\text{TL},j}^{(\ell)}(s)$ (5.26) involves the corresponding partial spectral function $\varrho_j^{(\ell)}(\sigma)$ (5.4). Although at the higher loop levels the latter appears to be rather cumbersome (see Section 4.2 and Appendix B), the integration on the right-hand side of Eq. (5.26) can be explicitly performed as well. For example, at the two-loop level ($\ell = 2$) the first-order ($j = 1$) expansion function (5.26), which plays a role of the two-loop effective couplant in the timelike domain $a_{\text{TL}}^{(2)}(s)$, reads

$$A_{\text{TL},1}^{(2)}(s) = A_{\text{TL},1}^{(1)}(s) - B_1 \frac{\theta(s - m^2)}{\ln^2 w + \pi^2} \left[W(s) - A_{\text{TL},1}^{(1)}(s) \ln w + 1 \right], \quad (5.37)$$

where $A_{\text{TL},1}^{(1)}(s)$ stands for the one-loop timelike effective couplant (5.33),

$$W(s) = \frac{1}{2} \ln \left[\ln^2 w + \pi^2 \right], \quad (5.38)$$

$B_1 = \beta_1/\beta_0^2$, and $w = s/\Lambda^2$, see also paper [105]. In turn, the two-loop second-order ($\ell = 2$, $j = 2$) effective expansion function (5.26), which is the

---

4. Note that at the one-loop and two-loop levels the perturbative approximation of the function $R(s)$ (5.29) is identical to the result of naive continuation (3.64).

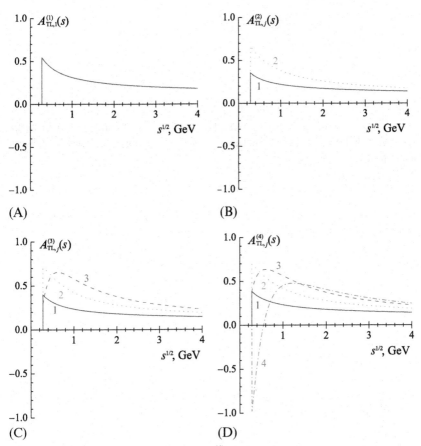

FIG. 5.7 The "timelike" expansion functions $A_{TL,j}^{(\ell)}(s)$ (5.26) at various loop levels ($1 \le \ell \le 4$, $1 \le j \le \ell$). *Plot A*: one-loop level ($\ell = 1$, $j = 1$). *Plot B*: two-loop level ($\ell = 2$, $1 \le j \le 2$), the function $A_{TL,2}^{(2)}(s)$ is scaled by the factor of 10. *Plot C*: three-loop level ($\ell = 3$, $1 \le j \le 3$), the functions $A_{TL,2}^{(3)}(s)$ and $A_{TL,3}^{(3)}(s)$ are scaled by the factors of 10 and 10^2, respectively. *Plot D*: four-loop level ($\ell = 4$, $1 \le j \le 4$), the functions $A_{TL,2}^{(4)}(s)$, $A_{TL,3}^{(4)}(s)$, and $A_{TL,4}^{(4)}(s)$ are scaled by the factors of 10, 10^2, and 10^3, respectively. *Numerical labels* specify the order j of the ℓ-loop function $A_{TL,j}^{(\ell)}(s)$. The values of parameters: $n_f = 3$ active flavors, $\Lambda = 350$ MeV.

continuation of the two-loop perturbative couplant squared $[a_s^{(2)}(Q^2)]^2$ into the timelike domain, can be represented in the following form:

$$A_{TL,2}^{(2)}(s) = \frac{\theta(s - m^2)}{\ln^2 w + \pi^2} + B_1 \frac{\theta(s - m^2)}{\left(\ln^2 w + \pi^2\right)^2} \left\{ A_{TL,1}^{(1)}(s) \left(\ln^2 w - \pi^2\right) \right.$$
$$\left. - \ln w [2W(s) + 1] \right\} + B_1^2 \frac{\theta(s - m^2)}{\left(\ln^2 w + \pi^2\right)^3} \left\{ \left(\ln^2 w - \frac{\pi^2}{3}\right) \right.$$

$$\times \left[\left(W(s) + \frac{1}{3} \right)^2 + \frac{1}{9} - \pi^2 \left[A_{\mathrm{TL},1}^{(1)}(s) \right]^2 \right]$$

$$\left. - \frac{2}{3} A_{\mathrm{TL},1}^{(1)}(s) \ln w \left[\ln^2 w - 3\pi^2 \right] \left(W(s) + \frac{1}{3} \right) \right\}. \tag{5.39}$$

Obviously, this function differs from the square of the two-loop first-order function (5.37). It is necessary to emphasize that, in general, the higher-order effective timelike expansion functions (5.26) differ from the respective powers of the first-order one, namely

$$A_{\mathrm{TL},j}^{(\ell)}(s) \neq \left[A_{\mathrm{TL},1}^{(\ell)}(s) \right]^j, \quad j \geq 2. \tag{5.40}$$

In turn, this implies that the strong correction to the R-ratio (5.25) constitutes an expansion over the set of functions (5.26) rather than a power series. As for the higher loop levels ($\ell > 2$), the integration in Eq. (5.26) can be performed either explicitly or numerically in the way specified in Appendix B.

The plots of the timelike expansion functions $A_{\mathrm{TL},j}^{(\ell)}(s)$ (5.26) up to the four-loop level ($1 \leq \ell \leq 4$, $1 \leq j \leq \ell$) are presented in Fig. 5.7. As one can infer from this figure, the expansion functions $A_{\mathrm{TL},j}^{(\ell)}(s)$ (5.26) contain no unphysical singularities and vanish in the kinematically forbidden interval $0 \leq s < m^2$. As it will be elucidated in Chapter 6 the effective expansion functions $A_{\mathrm{TL},j}^{(\ell)}(s)$ (5.26) incorporate the effects due to continuation of the spacelike theoretical results into the timelike domain and, therefore, differ from the corresponding result of naive continuation $[a_{\mathrm{s}}^{(\ell)}(|s|)]^j$ (3.64) starting from the three-loop level even in the ultraviolet asymptotic $s \to \infty$. At the same time at any given loop level the function $R(s)$ (5.27), which is composed of the expansion functions $A_{\mathrm{TL},j}^{(\ell)}(s)$ (5.26), coincides (up to the terms uncontrollable at the loop level on hand) with corresponding perturbative approximation of R-ratio (5.29) at high energies.[5] The expansion functions of first two orders ($j = 1, 2$) assume positive values, whereas the higher-order functions ($j \geq 3$) may also assume negative values. As discussed earlier the suppression of the involved higher-order partial spectral functions $\varrho_j^{(\ell)}(\sigma)$ (5.4) in the entire range $0 \leq \sigma < \infty$ eventually results in the subdominance of the higher-order expansion functions $A_{\mathrm{TL},j}^{(\ell)}(s)$ ($j \geq 2$) to the functions of the preceding orders. In particular, Fig. 5.7D signifies that the timelike expansion function of the $(j+1)$th order $A_{\mathrm{TL},j+1}^{(\ell)}(s)$ is suppressed by an order of magnitude with respect to the function of the jth order $A_{\mathrm{TL},j}^{(\ell)}(s)$. It is worthwhile to outline

---

5. In fact, the perturbative approximation of R-ratio of electron-positron annihilation into hadrons (5.29) represents the reexpansion of the expression (5.27) at high energies, see Chapter 6 for the details.

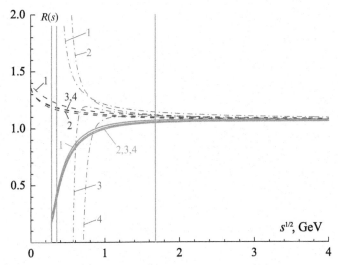

FIG. 5.8 The function $R(s)$ at various loop levels ($1 \leq \ell \leq 4$): DPT expression (Eq. 5.27, *solid curves*), its massless limit (Eq. 5.28, *dashed curves*), and the perturbative approximation (Eq. 5.29, *dot-dashed curves*). *Numerical labels* specify the loop level. *Vertical solid lines* mark the locations of (from left to right) the kinematic threshold at $\sqrt{s} = m$, the one-loop perturbative pole at $\sqrt{s} = \Lambda$, and the boundary of the convergence range of the perturbative approximation (5.29) at $\sqrt{s} = \exp(\pi/2)\Lambda$. The values of parameters: $n_f = 3$ active flavors, $\Lambda = 350$ MeV.

that for the perturbative power series (5.29) the suppression of the higher-order expansion functions $[a_s^{(\ell)}(|s|)]^j$ takes place at high energies only, whereas for the expression (5.27) the higher-order expansion functions $A_{\mathrm{TL},j}^{(\ell)}(s)$ are suppressed in the entire energy range. In turn, this fact leads to an enhanced stability of the function $R(s)$ calculated in the framework of dispersive approach (with respect to that of the QCD perturbation theory) to the higher loop corrections, and to the choice of the subtraction scheme at moderate and low energies.

In particular, Fig. 5.8 illustrates the variability of the function $R(s)$ calculated in the framework of various approaches due to the incorporation of the higher loop corrections. The results presented in this figure testify to an enhanced higher loop stability of the DPT expression for $R(s)$ (Eq. 5.27, solid curves) and its massless limit (Eq. 5.28, dashed curves) with respect to that of the perturbative approximation (Eq. 5.29, dot-dashed curves). Specifically, as one can infer from Fig. 5.8 at low and intermediate energies (apart from the aforementioned qualitative distinctions between the approaches on hand) the deviation between the curves corresponding to various loop levels ($1 \leq \ell \leq 4$) is significantly less for the DPT expression (5.27) and its massless limit (5.28) than for the perturbative approximation (5.29). As mentioned earlier this fact originates in the suppression of the higher-order expansion functions $A_{\mathrm{TL},j}^{(\ell)}(s)$ (5.26) with respect to those of the preceding orders in the entire energy range, as well as

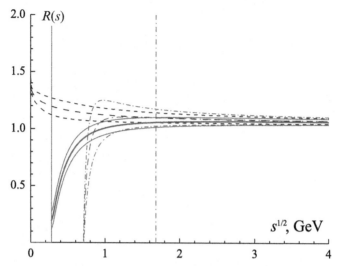

FIG. 5.9   The four-loop function $R(s)$ in the "MS-like" subtraction scheme: DPT expression (Eq. 5.27, *solid curves*), its massless limit (Eq. 5.28, *dashed curves*), and the perturbative approximation (Eq. 5.29, *dot-dashed curves*). *Thick curves* denote the results corresponding to the $\overline{\text{MS}}$-scheme, whereas *thin curves* portray the variation of the involved scheme-dependent coefficients. *Vertical solid line* marks the location of the kinematic threshold at $\sqrt{s} = m$, whereas *vertical dot-dashed line* marks the boundary of the convergence range of the approximation (5.29) at $\sqrt{s} = \exp(\pi/2)\Lambda$. The values of parameters: $n_f = 3$ active flavors, $\Lambda = 350\,\text{MeV}$.

in the presence of the unphysical singularities in the perturbative approximation of the function $R(s)$ (5.29), which makes its higher-order terms amplified at low energies. At the same time in the ultraviolet asymptotic all three expressions for the function $R(s)$ coincide within the accuracy of a given loop level.

The stability of the results obtained in the framework of the approaches on hand with respect to the choice of the subtraction scheme is illustrated by Fig. 5.9. Specifically, this figure displays the four-loop expression for the function $R(s)$ calculated within DPT (Eq. (5.27), solid curves), its massless limit (Eq. 5.28, dashed curves), and perturbative approximation (Eq. 5.29, dot-dashed curves). The internal (thick) curves correspond to the $\overline{\text{MS}}$-scheme, whereas the external (thin) curves mark the band corresponding to a probe variation, in the "MS-like" scheme, of the involved scheme-dependent coefficients (namely, d_j for $j = 2, 3, 4$ and β_j for $j = 2, 3$) within a certain range, the latter being the same for all three approaches. As one can infer from Fig. 5.9 at low and moderate energies the DPT and APT expressions for the function $R(s)$, Eqs. (5.27), (5.28), respectively, possess a higher stability to the variation of the involved scheme-dependent coefficients than that of the perturbative approximation (5.29). This feature is primarily caused by the fact that the perturbative approximation of the R-ratio (5.29) contains infrared unphysical singularities, that enhances its higher-order terms at low energies, whereas the higher-order terms of the DPT and APT expressions appear to be subleading to the terms of the preceding orders in the entire energy range. At the same time in the ultraviolet asymptotic $s \to \infty$ all three approaches provide the same expression for the function $R(s)$ (up to the terms uncontrollable at a given loop level), see also the discussion of the resummation of the π^2-terms presented in Chapter 6.

As discussed in Section 4.2, the employed spectral function $\rho_{\text{pert}}(\sigma)$ (4.55) retains only perturbative contributions and thereby brings only partial information about the functions on hand into the integral representations (4.34)–(4.36). Eventually, this results in a smooth behavior of the function $R(s)$ without resonance peaks, which makes its direct juxtaposition with relevant experimental data somewhat puzzling at low and intermediate energies. To obviate this difficulty one can additionally take into account the contributions responsible for the resonance structure of $R(s)$ in the spectral function $\rho(\sigma)$ by making use of, for example, Breit-Wigner parameterization (4.57) discussed in Section 4.2. At the same time one can alternatively resort to the widely employed method of "smearing" the function $R(s)$, which was proposed in mid-1970s by Poggio, Quinn, and Weinberg [102]. The basic idea of this method is quite simple and consists in shifting the points $(s + i\varepsilon)$ and $(s - i\varepsilon)$ in the definition (4.4) away[6] from the edges of physical cut of the hadronic vacuum polarization function $\Pi(q^2)$, specifically,

$$\bar{R}(s, \epsilon) = \frac{1}{2\pi i} \left[\Pi(s + i\epsilon) - \Pi(s - i\epsilon)\right], \qquad (5.41)$$

which makes the resonance structure of $R(s)$ less pronounced. The measure of the smoothness of the resulting function $\bar{R}(s, \epsilon)$ (5.41) is determined by the parameter ϵ of the dimension of GeV^2 ($\epsilon > 0$). Namely, for small values of ϵ the resonance peaks are kept visible and distinguishable from each other, whereas for large values of ϵ the resonance structure of the function $R(s)$ is averaged out.

In turn, the dispersion relation (4.1) makes it possible to express the smeared function $\bar{R}(s, \epsilon)$ (5.41) in terms of the original R-ratio (4.4), namely [102],

$$\bar{R}(s, \epsilon) = \frac{1}{2\pi i} (s + i\epsilon - s + i\epsilon) \int_{m^2}^{\infty} \frac{R(\zeta)}{(\zeta - s - i\epsilon)(\zeta - s + i\epsilon)} d\zeta$$

$$= \frac{\epsilon}{\pi} \int_{m^2}^{\infty} \frac{R(\zeta)}{(\zeta - s)^2 + \epsilon^2} d\zeta. \qquad (5.42)$$

It is worthwhile to mention here that the finiteness of the parameter ϵ in the definition (5.41) makes the smeared function $\bar{R}(s, \epsilon)$ nonvanishing[7] for

---

6. In other words, to assign a finite positive value ϵ to the parameter ε in the definition (4.4).

7. Contrary to the original function $R(s)$ (4.4).

the energies below the hadronic production threshold $s < m^2$. To obtain the experimental prediction for the smeared function $\bar{R}(s, \epsilon)$ one can follow the very same lines as in Sections 1.2 and 1.3. Specifically, for this purpose the R-ratio in the integrand of Eq. (5.42) can be approximated by the relevant experimental data at low and moderate energies and by its perturbative approximation at high energies, namely (1.42),

$$\bar{R}_{\text{exp}}(s, \epsilon) = \frac{\epsilon}{\pi} \int_{m^2}^{s_0} \frac{R_{\text{data}}(\zeta)}{(\zeta - s)^2 + \epsilon^2} d\zeta + \frac{\epsilon}{\pi} \int_{s_0}^{\infty} \frac{R_{\text{pert}}^{(\ell)}(\zeta)}{(\zeta - s)^2 + \epsilon^2} d\zeta. \tag{5.43}$$

In this equation $R_{\text{data}}(s)$ stands for the experimentally measured numerical values of the R-ratio of electron-positron annihilation into hadrons, $R_{\text{pert}}^{(\ell)}(s)$ denotes the perturbative approximation (5.29), and the parameter s_0 of the dimension of GeV^2 specifies the energy scale of "matching" of the quantities $R_{\text{data}}(s)$ and $R_{\text{pert}}^{(\ell)}(s)$. It has to be noted that the perturbative approximation of R-ratio (5.29) is applicable at rather high energies only. In particular, as it will be discussed in the next chapter, Eq. (5.29) is invalid for $s < \exp(\pi/2)\Lambda \simeq 4.81\Lambda$. In turn, this implies that either the value of the parameter s_0 in Eq. (5.43) has to be taken large enough to securely make $R_{\text{pert}}^{(\ell)}(s)$ applicable for any $s \geq s_0$ or in the integrand of the second term of Eq. (5.43) the perturbative approximation $R_{\text{pert}}^{(\ell)}(s)$ has to be replaced by $R^{(\ell)}(s)$ (5.27).

In the framework of the dispersively improved perturbation theory the smeared function $\bar{R}(s, \epsilon)$ (5.41) acquires the following form

$$\bar{R}(s, \epsilon) = \Delta \Pi^{(0)}(s + i\epsilon, s - i\epsilon) + \bar{r}(s, \epsilon). \tag{5.44}$$

In this equation the leading-order term, which is determined by Eq. (5.2), can be calculated numerically, whereas the strong correction $\bar{r}(s, \epsilon)$ can be expressed in terms of the spectral density $\rho(\sigma)$ (4.37) by making use of the integration by parts. Specifically, Eqs. (5.22), (5.42) imply that

$$\bar{r}(s, \epsilon) = \frac{\epsilon}{\pi} \int_{m^2}^{\infty} \frac{r(\zeta)}{(\zeta - s)^2 + \epsilon^2} d\zeta$$

$$= \frac{1}{\pi} \arctan\left(\frac{\zeta - s}{\epsilon}\right) r(\zeta) \Big|_{\zeta = m^2}^{\zeta = \infty} + \frac{1}{\pi} \int_{m^2}^{\infty} \arctan\left(\frac{\sigma - s}{\epsilon}\right) \rho(\sigma) \frac{d\sigma}{\sigma}$$

$$= \frac{1}{\pi} \int_{m^2}^{\infty} \left[\arctan\left(\frac{s - m^2}{\epsilon}\right) + \arctan\left(\frac{\sigma - s}{\epsilon}\right) \right] \rho(\sigma) \frac{d\sigma}{\sigma} \tag{5.45}$$

(for technical reasons it is preferable to keep the term in the squared brackets in the third line of this equation in its present form instead of reducing it to an inverse tangent of a single argument). It is also worth mentioning that in the

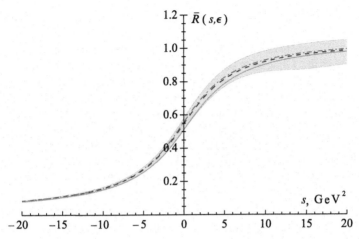

FIG. 5.10 The smeared function $\bar{R}(s, \epsilon)$ (5.41): experimental prediction (Eq. 5.43, *shaded band*), the DPT expression (Eq. 5.44, *solid curve*), its massless limit (Eq. 5.47, *dashed curve*), and the perturbative approximation (Eq. 5.48, *dot-dashed curve*). The presented results correspond to the four-loop level. The values of parameters: $\epsilon = 5\,\mathrm{GeV}^2$, $n_f = 3$ active flavors, $\Lambda = 350\,\mathrm{MeV}$.

limit $\epsilon \to 0_+$ the function $\bar{r}(s, \epsilon)$ (5.45) evidently coincides with the original strong correction $r(s)$ (4.18), namely,

$$\lim_{\epsilon \to 0_+} \bar{r}(s, \epsilon) = \int_{m^2}^{\infty} \left[\theta(s - m^2) - \theta(s - \sigma)\right] \rho(\sigma) \frac{d\sigma}{\sigma} = \theta(s - m^2) \int_{s}^{\infty} \rho(\sigma) \frac{d\sigma}{\sigma},$$
(5.46)

where $\theta(x)$ denotes the Heaviside unit step function (1.27).

In turn, in the framework of the analytic perturbation theory the smeared function $\bar{R}(s, \epsilon)$ (5.41) can be calculated in the very same way as within DPT, which eventually leads to

$$\bar{R}_{\mathrm{APT}}(s, \epsilon) = \frac{1}{2} + \frac{1}{\pi} \arctan\left(\frac{s}{\epsilon}\right) + \frac{1}{\pi} \int_{0}^{\infty} \left[\arctan\left(\frac{s}{\epsilon}\right) + \arctan\left(\frac{\sigma - s}{\epsilon}\right)\right] \rho(\sigma) \frac{d\sigma}{\sigma}.$$
(5.47)

Obviously, this equation coincides with the massless limit ($m = 0$) of the DPT expression (5.44). As for the calculation of the smeared function $\bar{R}(s, \epsilon)$ in the framework of perturbative approach, for this purpose one can directly employ the perturbative approximation of the hadronic vacuum polarization function in the definition (5.41), specifically

$$\bar{R}_{\mathrm{pert}}(s, \epsilon) = \frac{1}{2} + \frac{1}{\pi} \arctan\left(\frac{s}{\epsilon}\right) + \frac{1}{2\pi i}\left[p_{\mathrm{pert}}(s + i\epsilon) - p_{\mathrm{pert}}(s - i\epsilon)\right], \quad (5.48)$$

where $p_{\text{pert}}(q^2)$ stands for the perturbative strong correction to the hadronic vacuum polarization function, see Section 3.1.

Fig. 5.10 presents the comparison of the smeared function $\bar{R}(s, \epsilon)$ (5.41) calculated in the framework of the approaches on hand with its experimental prediction. In particular, as one can infer from this figure the DPT (Eq. 5.44, solid curve), APT (Eq. 5.47, dashed curve), and perturbative (Eq. 5.48, dot-dashed curve) expressions for the function $\bar{R}(s, \epsilon)$ possess a similar qualitative behavior and appear to be in a good agreement with each other as well as with the pertinent experimental prediction (Eq. 5.43, shaded band) for all values of s.

Thus the continuation of the spacelike theoretical results into the timelike domain (4.6) leads to an essential distortion of the original perturbative power series. In particular, the function $R(s)$ (5.27) constitutes an expansion over the set of functions $A^{(\ell)}_{\text{TL},j}(s)$ (5.26), which obviously differs from the naive continuation of the perturbative expression (3.57) into the timelike domain (3.64). As it will be discussed in detail in Chapter 6 even in the ultraviolet asymptotic $s \to \infty$ the resulting function $R(s)$ (5.27) deviates from its naive form (3.64) due to the appearance of the so-called π^2-terms. The latter nearly supersede the respective perturbative contributions and thereby play a substantial role in the study of the R-ratio of electron-positron annihilation into hadrons and related quantities in the entire energy range.

5.3 ADLER FUNCTION WITHIN DPT

As outlined in Chapter 4 the dispersive approach to QCD provides the following integral representation for the Adler function (4.36):

$$D(Q^2) = D^{(0)}(Q^2) + \frac{Q^2}{Q^2 + m^2} \int_{m^2}^{\infty} \rho(\sigma) \frac{\sigma - m^2}{\sigma + Q^2} \frac{d\sigma}{\sigma}, \tag{5.49}$$

which contains no unphysical singularities and properly accounts for all the nonperturbative constraints discussed in Section 4.1. In Eq. (5.49) the leading-order term reads (4.41)

$$D^{(0)}(Q^2) = 1 + \frac{3}{\xi} \left[1 - \sqrt{1 + \xi^{-1}} \operatorname{arcsinh}\left(\xi^{1/2}\right) \right], \tag{5.50}$$

where $\xi = Q^2/m^2$, $Q^2 = -q^2 \geq 0$ is the spacelike kinematic variable, m denotes the total mass of the lightest allowed hadronic final state, and $\rho(\sigma)$ is the spectral density (4.37), which brings the corresponding perturbative input into the integral representation (5.49). In particular, as discussed in Section 4.2 and Appendix B, the perturbative part of the spectral density (4.55) can be explicitly calculated at an arbitrary loop level, specifically,

$$\rho^{(\ell)}_{\text{pert}}(\sigma) = \sum_{j=1}^{\ell} d_j \varrho^{(\ell)}_j(\sigma). \tag{5.51}$$

In this equation $\varrho_j^{(\ell)}(\sigma)$ is the partial spectral function (5.4), which corresponds to the jth power of the ℓ-loop perturbative couplant, and d_j stand for the coefficients of the Adler function perturbative expansion (see Section 3.2). In general, at any given loop level the integration on the right-hand side of Eq. (5.49) can be performed for the whole spectral function (5.51) in the way described in Appendix B. At the same time for illustrative purposes it is worthwhile to handle each term of the sum (5.51) separately from the other terms. In particular, in this case the ℓ-loop strong correction to the Adler function acquires the form

$$d^{(\ell)}(Q^2) = \sum_{j=1}^{\ell} d_j A_{\text{SL},j}^{(\ell)}(Q^2), \qquad (5.52)$$

which makes it possible to clearly distinguish term-by-term the respective perturbative contributions to the integral representation (5.49). In Eq. (5.52)

$$A_{\text{SL},j}^{(\ell)}(Q^2) = \frac{Q^2}{Q^2 + m^2} \int_{m^2}^{\infty} \varrho_j^{(\ell)}(\sigma) \frac{\sigma - m^2}{\sigma + Q^2} \frac{d\sigma}{\sigma} \qquad (5.53)$$

is the ℓ-loop effective "spacelike" expansion function of jth order, which plays the role of jth power of ℓ-loop perturbative couplant $[a_s^{(\ell)}(Q^2)]^j$. Evidently, the functions $A_{\text{SL},j}^{(\ell)}(Q^2)$ (5.53) differ from the expansion functions $A_{\text{HP},j}^{(\ell)}(Q^2)$ (5.12) and $A_{\text{TL},j}^{(\ell)}(s)$ (5.26) discussed in Sections 5.1 and 5.2, respectively. As one may note the spacelike expansion function $A_{\text{SL},j}^{(\ell)}(Q^2)$ (5.53) can be obtained as the logarithmic derivative of $A_{\text{HP},j}^{(\ell)}(Q^2)$ (5.12), namely,

$$A_{\text{SL},j}^{(\ell)}(Q^2) = \frac{d}{d\ln Q^2} A_{\text{HP},j}^{(\ell)}(Q^2). \qquad (5.54)$$

Despite the form of the strong correction $d^{(\ell)}(Q^2)$ (5.52) formally resembles its perturbative approximation (3.57) and in the ultraviolet asymptotic $Q^2 \to \infty$ Eqs. (5.52), (3.57) coincide, at low energies the functions $d^{(\ell)}(Q^2)$ (5.52) and $d_{\text{pert}}^{(\ell)}(Q^2)$ (3.57) appear to be rather different. Specifically, as it will be discussed later, the jth order expansion function $A_{\text{SL},j}^{(\ell)}(Q^2)$ ($j \geq 2$) differs from the jth power of the first-order function $[A_{\text{SL},1}^{(\ell)}(Q^2)]^j$ and, hence, the strong correction (5.52) constitutes an expansion over the set of functions (5.53) rather than a power series.

Thus in the framework of dispersively improved perturbation theory the Adler function at the ℓ-loop level reads

$$D^{(\ell)}(Q^2) = D^{(0)}(Q^2) + \sum_{j=1}^{\ell} d_j A_{\text{SL},j}^{(\ell)}(Q^2), \qquad (5.55)$$

where the leading-order term is specified in Eq. (5.50) and the spacelike effective expansion functions $A_{\text{SL},j}^{(\ell)}(Q^2)$ are given in Eq. (5.53). It is worth mentioning that in the massless limit ($m = 0$) the DPT expression for $D(Q^2)$ (5.55) becomes identical to that of the analytic perturbation theory (see Section 4.3)

$$D_{\text{APT}}^{(\ell)}(Q^2) = 1 + \sum_{j=1}^{\ell} d_j \int_0^{\infty} \frac{\varrho_j^{(\ell)}(\sigma)}{\sigma + Q^2} d\sigma. \tag{5.56}$$

At the same time in the ultraviolet asymptotic the Adler function can be approximated by the perturbative power series (see Section 3.2)

$$D_{\text{pert}}^{(\ell)}(Q^2) = 1 + \sum_{j=1}^{\ell} d_j \left[a_{\text{s}}^{(\ell)}(Q^2) \right]^j, \quad Q^2 \to \infty, \tag{5.57}$$

where $a_{\text{s}}^{(\ell)}(Q^2)$ stands for the ℓ-loop perturbative couplant (see Chapter 2 and Appendix A).

Let us first address the Adler function $D(Q^2)$ obtained in the framework of the dispersively improved perturbation theory (5.55) at the one-loop level ($\ell = 1$). In this case the corresponding strong correction (5.52) consists of only one term, so that

$$D^{(1)}(Q^2) = D^{(0)}(Q^2) + d_1 A_{\text{SL},1}^{(1)}(Q^2). \tag{5.58}$$

In this equation the leading-order term is given in Eq. (5.50), $d_1 = 4/\beta_0$ and $\beta_0 = 11 - 2n_f/3$ denote the one-loop perturbative coefficients of the Adler function (see Section 3.2) and β function (see Section 2.1), respectively, n_f stands for the number of active flavors,

$$A_{\text{SL},1}^{(1)}(Q^2) = \frac{Q^2}{Q^2 + m^2} \int_{m^2}^{\infty} \frac{1}{\ln^2(\sigma/\Lambda^2) + \pi^2} \frac{\sigma - m^2}{\sigma + Q^2} \frac{d\sigma}{\sigma} \tag{5.59}$$

is the one-loop spacelike expansion function of the first order (5.53), and Λ denotes the QCD scale parameter. As discussed earlier the function $A_{\text{SL},1}^{(1)}(Q^2)$ (5.59) plays a role of the one-loop effective couplant $a_{\text{SL}}^{(1)}(Q^2)$ in the spacelike domain. The integration on the right-hand side of Eq. (5.59) can be performed numerically in the way described in Appendix B. At the same time, for the illustrative purposes one may also represent Eq. (5.59) in the following form:

$$A_{\text{SL},1}^{(1)}(Q^2) = \frac{1}{\ln z} + \frac{z}{1 - z} \frac{1 + \chi}{z + \chi} - \frac{z}{z + \chi} \int_0^{\chi} \frac{1 - \chi/\sigma}{\ln^2 \sigma + \pi^2} \frac{d\sigma}{\sigma + z}, \tag{5.60}$$

where $z = Q^2/\Lambda^2$ and $\chi = m^2/\Lambda^2$. The first term on the right-hand side of this equation is the one-loop perturbative QCD couplant (2.19), the second term eliminates its unphysical singularity at $Q^2 = \Lambda^2$, whereas the third term constitutes an infrared correction. Evidently, in the massless limit ($m = 0$) the

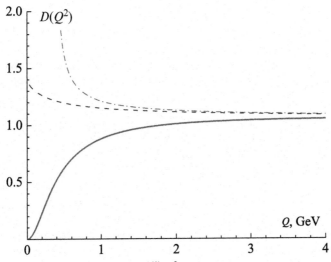

FIG. 5.11 The one-loop Adler function $D^{(1)}(Q^2)$: DPT expression (Eq. 5.58, *solid curve*), its massless limit (Eq. 5.61, *dashed curve*), and the perturbative approximation (Eq. 5.65, *dot-dashed curve*). The values of parameters: $n_f = 3$ active flavors, $\Lambda = 350$ MeV.

DPT expression (5.58) becomes identical to the APT prediction for the Adler function (5.56), namely (see also discussion in Section 4.3),

$$D_{APT}^{(1)}(Q^2) = 1 + d_1 a_{APT}^{(1)}(Q^2), \tag{5.61}$$

where

$$a_{APT}^{(1)}(Q^2) = \int_0^\infty \frac{1}{\ln^2(\sigma/\Lambda^2) + \pi^2} \frac{d\sigma}{\sigma + Q^2} = \frac{1}{\ln z} + \frac{1}{1 - z}, \quad z = \frac{Q^2}{\Lambda^2} \tag{5.62}$$

stands for the one-loop infrared finite analytic couplant[8] [353, 354]. Note that the function (5.62) can also be expressed in terms of the one-loop perturbative couplant (2.19) $a_s^{(1)}(Q^2) = 1/\ln(Q^2/\Lambda^2)$, specifically,

$$a_{APT}^{(1)}(Q^2) = a_s^{(1)}(Q^2) + \frac{1}{1 - \exp\left[1/a_s^{(1)}(Q^2)\right]}. \tag{5.63}$$

As discussed in Section 2.2, the first term on the right-hand side of this equation contains unphysical singularity at $Q^2 = \Lambda^2$, which is subtracted, in a minimal way, by the second term. The latter, being of an intrinsically nonperturbative nature, cannot be described in the framework of perturbation theory. In particular, the function

---

8. The expression (5.62) was also presented in the paper [421].

$$\frac{1}{1 - \exp{(1/a)}} \qquad (5.64)$$

cannot be approximated by the power series in a at $a \to 0_+$, because all the coefficients of its Taylor expansion at $a \to 0_+$ vanish identically.[9] As for the perturbative approximation of the Adler function (5.57), at the one-loop level it reads

$$D^{(1)}_{\text{pert}}(Q^2) = 1 + d_1 \frac{1}{\ln(Q^2/\Lambda^2)}, \qquad (5.65)$$

see Section 3.2 for the details.

The plots of the one-loop Adler function $D^{(1)}(Q^2)$ calculated within approaches on hand are displayed in Fig. 5.11. As one can infer from this figure the perturbative approximation of the Adler function (Eq. 5.65, dot-dashed curve) diverges in the infrared domain due to the unphysical singularities, that makes it inapplicable at low energies. In turn, the analytic perturbation theory, on the one hand, eliminates such intrinsic difficulties of the perturbative approach as the infrared unphysical singularities and somewhat extends its range of applicability toward the low energies. However, on the other hand, in the infrared domain the qualitative behavior of the APT prediction for the Adler function (Eq. 5.61, dashed curve) appears to be experimentally disfavored. At the same time as one can infer from Fig. 5.11, the dispersively improved perturbation theory provides the Adler function (Eq. 5.58, solid curve), which contains no unphysical singularities and possesses physically sound behavior in the entire energy range.

As discussed earlier, a rigorous theoretical method, which would have enabled one to thoroughly reckon all the intricate nonperturbative QCD effects, is rather far from being feasible. In turn, this fact implies that the perturbation theory actually still remains the basic tool commonly employed in a variety of QCD applications. The QCD invariant charge $\alpha_s(Q^2)$ is defined as the product of the renormalized vertex and the respective powers of the corresponding Green's functions. However, as noted in Chapter 2, the implementation of this definition at a given order of perturbation theory results in the perturbative QCD running coupling, which contains infrared unphysical singularities, that, in turn, makes it useless in the study of the strong interaction processes at low energies. In general, to construct an effective extension of the perturbative strong running coupling to the infrared domain, additional information is obviously required. One of the simplest examples of such synthesis is the well-known Richardson's potential model. The latter was proposed in late 1970s in the paper [27] and has been successfully applied in various issues of the meson spectroscopy. The basic idea of that model was to merge the high-energy perturbative behavior of the QCD running coupling with the low-energy nonperturbative constraint

---

9. In other words, the function (5.64) is flat at $a \to 0_+$.

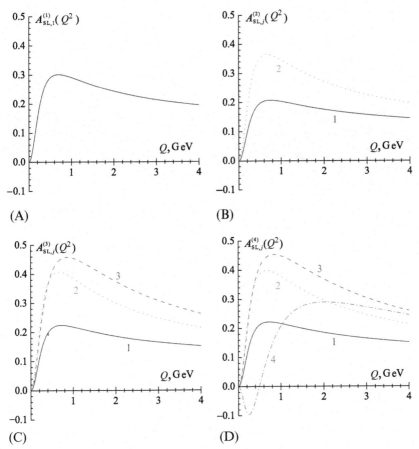

FIG. 5.12 The "spacelike" effective expansion functions $A_{\mathrm{SL},j}^{(\ell)}(Q^2)$ (5.53) at various loop levels ($1 \leq \ell \leq 4$, $1 \leq j \leq \ell$). *Plot A*: one-loop level ($\ell = 1, j = 1$). *Plot B*: two-loop level ($\ell = 2$, $1 \leq j \leq 2$), the function $A_{\mathrm{SL},2}^{(2)}(Q^2)$ is scaled by the factor of 10. *Plot C*: three-loop level ($\ell = 3$, $1 \leq j \leq 3$), the functions $A_{\mathrm{SL},2}^{(3)}(Q^2)$ and $A_{\mathrm{SL},3}^{(3)}(Q^2)$ are scaled by the factors of 10 and 10^2, respectively. *Plot D*: four-loop level ($\ell = 4$, $1 \leq j \leq 4$), the functions $A_{\mathrm{SL},2}^{(4)}(Q^2)$, $A_{\mathrm{SL},3}^{(4)}(Q^2)$, and $A_{\mathrm{SL},4}^{(4)}(Q^2)$ are scaled by the factors of 10, 10^2, and 10^3, respectively. *Numerical labels* indicate the order j of the ℓ-loop function $A_{\mathrm{SL},j}^{(\ell)}(Q^2)$. The values of parameters: $n_f = 3$ active flavors, $\Lambda = 350$ MeV.

prescribed by the concept of linear confinement in the one-gluon-exchange approximation (see also reviews [30–32] and references therein for the details). At the same time such method of construction of an effective extension of the perturbative strong running coupling to the infrared domain is certainly model-dependent. First of all this is so because the pattern of physical realizations of the nonperturbative dynamics of colored fields is quite diverse. Hence, the infrared behavior of the effective QCD running coupling, in general, depends

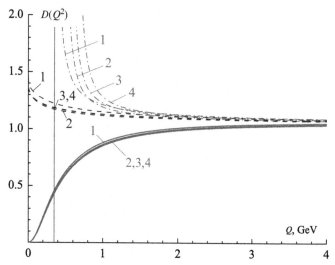

FIG. 5.13 The Adler function $D^{(\ell)}(Q^2)$ at various loop levels ($1 \le \ell \le 4$): DPT expression (Eq. 5.55, *solid curves*), its massless limit (Eq. 5.56, *dashed curves*), and the perturbative approximation (Eq. 5.57, *dot-dashed curves*). *Numerical labels* specify the loop level, whereas the *vertical solid line* marks the location of the one-loop perturbative pole at $Q = \Lambda$. The values of parameters: $n_f = 3$ active flavors, $\Lambda = 350$ MeV.

on a particular nonperturbative aspect of the strong interactions selected to be merged with the corresponding perturbative input, as well as on the way how this merge is accomplished. For example, the well-known Kugo–Ojima color confinement criterion, which was proposed in late 1970s in paper [422] (see also Refs. [423–427]), basically relies on the same physical grounds as the Richardson's approach does, namely, on the confinement of colored fields. However, the way the latter is accounted for (or modeled) within Richardson's framework differs from that of the Kugo–Ojima one, which eventually leads to different low-energy behavior of the effective QCD running coupling within these two approaches (see, in particular, paper [27] and, e.g., Refs. [428, 429]).

Let us proceed now to the Adler function $D^{(\ell)}(Q^2)$ at the higher loop levels ($\ell \ge 2$). In this case the DPT strong correction $d^{(\ell)}(Q^2)$ (5.52) takes the form

$$d^{(\ell)}(Q^2) = d_1 A^{(\ell)}_{SL,1}(Q^2) + d_2 A^{(\ell)}_{SL,2}(Q^2) + d_3 A^{(\ell)}_{SL,3}(Q^2) + \cdots . \qquad (5.66)$$

The number of terms on the right-hand side of this equation is determined by the loop level. As mentioned earlier at the higher loop levels the partial spectral functions $\varrho_j^{(\ell)}(\sigma)$ (5.4), which appear in the definition of the ℓ-loop expansion functions $A^{(\ell)}_{SL,j}(Q^2)$ (5.53), become rather cumbrous, so that the integration in Eq. (5.53) can be performed only numerically (see Appendix B). The plots of

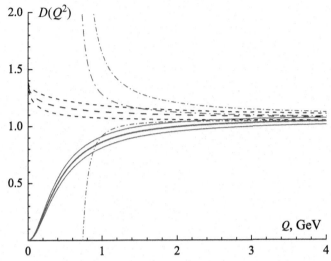

FIG. 5.14 The four-loop Adler function $D(Q^2)$ in the "$\overline{\text{MS}}$-like" subtraction scheme: DPT expression (Eq. 5.55, *solid curves*), its massless limit (Eq. 5.56, *dashed curves*), and the perturbative approximation (Eq. 5.57, *dot-dashed curves*). *Thick curves* denote the results corresponding to the $\overline{\text{MS}}$-scheme, whereas *thin curves* portray the variation of the involved scheme-dependent coefficients. The values of parameters: $n_f = 3$ active flavors, $\Lambda = 350\,\text{MeV}$.

the spacelike effective expansion functions $A^{(\ell)}_{\text{SL},j}(Q^2)$ (5.53) up to the four-loop level ($1 \leq \ell \leq 4$, $1 \leq j \leq \ell$) are presented in Fig. 5.12. As one can infer from this figure the functions $A^{(\ell)}_{\text{SL},j}(Q^2)$ (5.53) contain no unphysical singularities, vanish in the infrared limit $Q^2 \to 0$, and tend to the corresponding perturbative expressions $[a^{(\ell)}_s(Q^2)]^j$ at high energies. The spacelike expansion functions of the first two orders ($j = 1, 2$) assume positive values, whereas the higher-order functions ($j \geq 3$) may also assume negative values. It is necessary to emphasize here that the higher-order spacelike effective expansion functions (5.53) differ from the respective powers of the first-order one,[10] namely,

$$A^{(\ell)}_{\text{SL},j}(Q^2) \neq \left[A^{(\ell)}_{\text{SL},1}(Q^2)\right]^j, \quad j \geq 2, \tag{5.67}$$

which, in turn, implies that the strong correction to the Adler function (5.52) constitutes an expansion over the set of functions (5.53) rather than a power series. The higher-order expansion functions $A^{(\ell)}_{\text{SL},j}(Q^2)$ ($j \geq 2$) are suppressed with respect to the functions of preceding orders in the entire energy range $0 \leq Q^2 < \infty$, which, as discussed earlier, is caused by the subdominance of

---

10. Though, $A^{(\ell)}_{\text{SL},j}(Q^2) = [A^{(\ell)}_{\text{SL},1}(Q^2)]^j = [a^{(\ell)}_s(Q^2)]^j$ in the ultraviolet asymptotic $Q^2 \to \infty$.

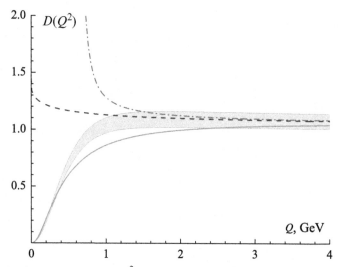

FIG. 5.15 The Adler function $D(Q^2)$: the experimental prediction (Eq. 1.52, *shaded band*), the DPT expression (Eq. 5.55, *solid curve*), its massless limit (Eq. 5.56, *dashed curve*), and the perturbative approximation (Eq. 5.57, *dot-dashed curve*). The presented results correspond to the four-loop level. The values of parameters: $n_f = 3$ active flavors, $\Lambda = 350$ MeV.

the involved higher-order partial spectral functions $\varrho_j^{(\ell)}(\sigma)$ (5.4) to those of the preceding orders in the entire range of their argument $0 \leq \sigma < \infty$ (see Section 4.2). In particular, as one can infer from Fig. 5.12D, the spacelike effective expansion function of the $(j+1)$th order $A_{\mathrm{SL},j+1}^{(\ell)}(Q^2)$ is suppressed by an order of magnitude with respect to the function of the jth order $A_{\mathrm{SL},j}^{(\ell)}(Q^2)$. It is also worth mentioning that in the framework of perturbation theory a similar suppression of the higher-order expansion functions $[a_{\mathrm{s}}^{(\ell)}(Q^2)]^j$ takes place in the ultraviolet domain only. In turn, this fact makes the Adler function $D(Q^2)$ calculated within dispersive approach more stable (with respect to that of calculated within QCD perturbation theory) to the higher loop corrections and the choice of the subtraction scheme at moderate and low energies.

In particular, the higher loop stability of the Adler function $D(Q^2)$ obtained in the framework of dispersively improved perturbation theory is illustrated by Fig. 5.13. The latter displays $D^{(\ell)}(Q^2)$ calculated at the first four loop levels ($1 \leq \ell \leq 4$) within the approaches on hand. As one can infer from this figure, at low energies the incorporation of the higher loop corrections to the Adler function calculated within the DPT approach (Eq. 5.55, solid curves) causes somewhat less variation of the resulting expression than that of calculated in the massless limit, which corresponds to the APT approach (Eq. 5.56, dashed curves), and substantially less than that of calculated in the framework of perturbative approach (Eq. 5.57, dot-dashed curves). As mentioned earlier such effect takes place due to the infrared divergence of the perturbative expansion functions $[a_s^{(\ell)}(Q^2)]^j$ at finite positive value of Q^2, which amplifies the respective higher-order contributions at low energies, whereas the higher-order expansion functions $A_{SL,j}^{(\ell)}(Q^2)$ (5.53) are suppressed with respect to those of the preceding orders in the entire energy range. At the same time at moderate energies the resulting expressions for the Adler function calculated within the approaches on hand tend to each other, whereas at high energies all three approaches yield the same result for the Adler function.

Fig. 5.14 illustrates the stability of the results obtained in the framework of the approaches on hand with respect to the choice of the subtraction scheme. In particular, this figure displays the four-loop Adler function $D(Q^2)$ calculated within dispersively improved perturbation theory (Eq. 5.55, solid curves), its massless limit (Eq. 5.56, dashed curves), and the perturbative approximation (Eq. 5.57, dot-dashed curves). The internal (thick) curves mark the results corresponding to the $\overline{MS}$-scheme, whereas the external (thin) curves delimitate the variability range of $D(Q^2)$ caused by a probe alternation of the involved scheme-dependent coefficients (specifically, d_j for $j = 2, 3, 4$ and β_j for $j = 2, 3$) within a certain range (the same for all three approaches), that corresponds to the "MS-like" scheme. As one can infer from Fig. 5.14, at low and intermediate energies the DPT expression for the Adler function (5.55) possesses a somewhat higher stability to the variation of the involved scheme-dependent coefficients than that of its massless limit (5.56). As for the perturbative approximation of the Adler function (5.57) its stability with respect to the choice of the subtraction scheme appears to be substantially less than those of the DPT and APT predictions. As discussed earlier such effect takes place due to the divergence of the perturbative expansion functions $[a_s^{(\ell)}(Q^2)]^j$ in the infrared domain, that makes the contributions of the scheme-dependent terms to the perturbative Adler function (5.57) amplified at low energies. At the same time the expressions for the Adler function calculated within approaches on hand tend to each other at moderate energies and coincide at high energies.

Fig. 5.15 displays the comparison of the Adler function calculated at the four-loop level in the framework of the approaches on hand with its experimental prediction obtained in Section 1.3. As one can infer from this figure the perturbative approximation of the Adler function (Eq. 5.57, dot-dashed curve) fails to describe $D(Q^2)$ at low energies. This is basically caused by such inherent difficulties of the QCD perturbation theory as the infrared unphysical singularities, which invalidate it at low energies. As for the APT prediction for the Adler function (Eq. 5.56, dashed curve), on the one hand, it eliminates the infrared unphysical singularities of the perturbative approach and somewhat extends its range of applicability toward the low energies. However, on the other hand, the APT approach fails to describe the Adler function in the infrared domain, that also makes it inapplicable at low energies. As discussed earlier this is primarily caused by the fact that the APT approach disregards the effects due to the nonvanishing hadronic production threshold, which play a valuable role in the infrared domain. At the same time as one can infer from Fig. 5.15, the dispersively improved perturbation theory provides the expression for the Adler function (Eq. 5.55, solid curve), which contains no unphysical singularities, proves to be in a good qualitative agreement with the experimental prediction for $D(Q^2)$ in the entire energy range, and thereby enables one to study a variety of the strong interaction processes beyond the limitations of the QCD perturbation theory.

Chapter 6

R-Ratio at High Energies

In general, to describe the *R*-ratio of electron-positron annihilation into hadrons in a self-consistent way, one has to apply the QCD perturbation theory together with respective dispersion relations. Eventually, this yields the function $R(s)$, which substantially differs from the naive continuation of the spacelike perturbative results into the timelike domain even at high energies. This is primarily caused by the appearance of the so-called π^2-terms, which dominate over the corresponding perturbative contributions and thereby play an essential role in the study of *R*-ratio and the related quantities in the entire energy range.

6.1 APPEARANCE OF THE π^2-TERMS

It is necessary to emphasize again that since the QCD perturbation theory is not directly applicable to the study of observables depending on the timelike kinematic variable, the only consistent way to calculate the *R*-ratio of electron-positron annihilation into hadrons is to employ the pertinent dispersion relations (4.1)–(4.6). Basically, in the ultraviolet asymptotic the nonperturbative effects become insignificant, so that the integral representation (4.6) together with perturbative input (3.57) yield the function $R(s)$, which is uniquely determined by the QCD perturbation theory and the effects due to continuation of its results into the timelike domain.

Let us examine now the high-energy behavior of the *R*-ratio in more detail. As noted in Section 5.2, the function $R(s)$ (5.27) is composed of the timelike effective expansion functions $A_{\mathrm{TL},j}^{(\ell)}(s)$ (5.26), which, in general, differ from the naive continuation of the respective terms of the perturbative power series into the timelike domain even in the ultraviolet asymptotic $s \to \infty$, namely,

$$A_{\mathrm{TL},j}^{(\ell)}(s) \neq \left[a_{\mathrm{s}}^{(\ell)}(|s|) \right]^{j}, \tag{6.1}$$

where $a_{\mathrm{s}}^{(\ell)}(Q^2)$ stands for the ℓ-loop perturbative couplant, see Chapter 2 and Appendix A. Nonetheless, as argued in Section 4.2, at an arbitrary loop level the reexpansion of Eq. (5.27) at $s \to \infty$ reproduces term-by-term the naive

Strong Interactions in Spacelike and Timelike Domains. http://dx.doi.org/10.1016/B978-0-12-803439-2.00006-5

expression (3.64) and additionally produces an infinite number of terms,[1] which are proportional to the integer powers of π^2. Basically, starting from the three-loop level some of these terms appear to be of the same order of magnitude as those of the respective perturbative expansion and, hence, must be properly accounted for even in the ultraviolet asymptotic. A discussion of this issue can also be found in, for example, papers [107, 108, 151, 430] and the references therein.

Let us first address the ultraviolet behavior of the function $R(s)$ at the one-loop level ($\ell = 1$). In this case, Eq. (5.27) acquires the following form[2]:

$$R^{(1)}(s) = 1 + d_1 A^{(1)}_{\text{TL},1}(s), \qquad (6.2)$$

where $d_1 = 4/\beta_0$ and $\beta_0 = 11 - 2n_f/3$ are the one-loop coefficients of perturbative expansions of the Adler function (see Section 3.2) and β function (see Section 2.1), respectively, n_f denotes the number of active flavors,

$$A^{(1)}_{\text{TL},1}(s) = \frac{1}{2} - \frac{1}{\pi} \arctan\left(\frac{\ln w}{\pi}\right), \qquad w = \frac{s}{\Lambda^2} \qquad (6.3)$$

is the one-loop effective timelike expansion function of the first order (5.26) in the massless limit, and Λ is the QCD scale parameter. Then given that for large values of x

$$\arctan x \simeq \frac{\pi}{2} - \frac{1}{x} + \frac{1}{3x^3} + \mathcal{O}\left(\frac{1}{x^5}\right), \qquad x \to \infty, \qquad (6.4)$$

in the ultraviolet asymptotic $s \to \infty$, Eq. (6.3) can be represented as

$$A^{(1)}_{\text{TL},1}(s) \simeq a^{(1)}_{\text{s}}(|s|) - \frac{\pi^2}{3}\frac{1}{\ln^3 w} + \frac{\pi^4}{5}\frac{1}{\ln^5 w} - \frac{\pi^6}{7}\frac{1}{\ln^7 w} + \mathcal{O}\left(\frac{1}{\ln^9 w}\right), \quad s \to \infty, \qquad (6.5)$$

with $a^{(1)}_{\text{s}}(Q^2) = 1/\ln(Q^2/\Lambda^2)$ being the one-loop perturbative couplant (2.19). Since the terms, which comprise the difference between the expressions $A^{(1)}_{\text{TL},1}(s)$ (6.3) and $a^{(1)}_{\text{s}}(|s|)$ (2.19) at high energies are beyond the accuracy of the one-loop level, the R-ratio of electron-positron annihilation into hadrons (6.2) can be approximated by its naive form (3.64), specifically,

$$R^{(1)}(s) \simeq 1 + d_1 a^{(1)}_{\text{s}}(|s|) + \mathcal{O}\left(\frac{1}{\ln^3 w}\right) = R^{(1)}_{\text{naive}}(s) + \mathcal{O}\left(\frac{1}{\ln^3 w}\right), \quad s \to \infty. \qquad (6.6)$$

This expression can be employed for the energies s, which assume that the terms of the order of $1/\ln^2(s/\Lambda^2)$ are negligible.

It also has to be mentioned that the reexpansion (6.5) diverges at low energies, whereas the original function (6.3) is free of the unphysical singularities.

---

1. In general, the terms of the reexpansion of the difference between the expressions $R(s)$ (5.27) and $R_{\text{naive}}(s)$ (3.64) in the ultraviolet asymptotic $s \to \infty$ are called "π^2-terms."
2. Throughout this chapter the limit $s \to \infty$ is assumed, so that the effects due to the nonvanishing value of the hadronic production threshold can be safely neglected.

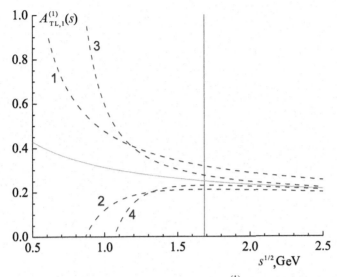

FIG. 6.1 One-loop timelike effective expansion function $A^{(1)}_{TL,1}(s)$ (Eq. 6.3, *solid curve*) and its approximations corresponding to various orders of the reexpansion (Eq. 6.5, *dashed curves*). *Numerical labels* specify the number of terms retained on the right-hand side of Eq. (6.5). *Vertical solid line* marks the boundary of the range of convergence of the reexpansion (6.5) at $\sqrt{s} = \exp(\pi/2)\Lambda$. The employed value of the QCD scale parameter is $\Lambda = 350$ MeV.

Additionally, the range of convergence of the series on the right-hand side of Eq. (6.4) is $x > 1$. Therefore the one-loop timelike effective expansion function $A^{(1)}_{TL,1}(s)$ (6.3) can be approximated by the sum of its naive form $a^{(1)}_s(|s|)$ and the respective π^2-terms (6.5) only for the energies $\sqrt{s} > \exp(\pi/2)\Lambda \simeq 4.81\Lambda$, whereas for the energies $\sqrt{s} < \exp(\pi/2)\Lambda$ the representation of the function (6.3) in the form of Eq. (6.5) becomes invalid. This issue is illustrated in Fig. 6.1, which displays the one-loop timelike effective couplant $A^{(1)}_{TL,1}(s)$ (Eq. 6.3, solid curve) and its approximations (dashed curves) corresponding to various orders of the reexpansion (6.5). Specifically, the first-order approximation, which retains only the first term on the right-hand side of Eq. (6.5), corresponds to the naive continuation of the one-loop perturbative couplant into the timelike domain given by Eq. (6.6). The higher-order approximations shown in Fig. 6.1 include first two, three, and four terms on the right-hand side of Eq. (6.5), respectively. As one can infer from this figure for $\sqrt{s} > \exp(\pi/2)\Lambda$ the reexpansion (6.5) converges[3] to the one-loop timelike effective couplant $A^{(1)}_{TL,1}(s)$ (6.3), whereas for $\sqrt{s} < \exp(\pi/2)\Lambda$ the expression (6.5) diverges. Also note that at the one-loop level one commonly truncates the reexpansion (6.5) at the first term thereby neglecting all the relevant π^2-terms. However, as one can infer from Fig. 6.1, the latter become nonnegligible when

---

3. Though, somewhat slowly when $\sqrt{s}$ approaches $\exp(\pi/2)\Lambda$.

the energy $\sqrt{s}$ approaches the value of $\exp(\pi/2)\Lambda$ (marked by the vertical solid line), which delimitates the range of convergence of the reexpansion (6.5). In particular, it turns out that even at moderate energies $a_s^{(1)}(|s|)$ considerably differs from $A_{\text{TL},1}^{(1)}(s)$, for example, at $\sqrt{s} = 2.5\,\text{GeV}$ $a_s^{(1)}(|s|)$ exceeds $A_{\text{TL},1}^{(1)}(s)$ by about 20%, see Fig. 6.1.

The incorporation of the two-loop corrections ($\ell = 2$) does not bring in any qualitative difference. Namely, in this case the function $R(s)$ (5.27) reads

$$R^{(2)}(s) = 1 + d_1 A_{\text{TL},1}^{(2)}(s) + d_2 A_{\text{TL},2}^{(2)}(s), \tag{6.7}$$

where the two-loop timelike effective expansion functions $A_{\text{TL},1}^{(2)}(s)$ and $A_{\text{TL},2}^{(2)}(s)$ are given by Eqs. (5.37), (5.39), respectively. In turn, at high energies these functions can be approximated by

$$A_{\text{TL},1}^{(2)}(s) \simeq a_s^{(2)}(|s|) - \frac{\pi^2}{3}\frac{1}{\ln^3 w} + \pi^2 \frac{B_1}{\ln^4 w}\left[\ln(\ln w) - \frac{5}{6}\right] + \mathcal{O}\left(\frac{1}{\ln^5 w}\right), \tag{6.8}$$

$$A_{\text{TL},2}^{(2)}(s) \simeq \left[a_s^{(2)}(|s|)\right]^2 - \frac{\pi^2}{\ln^4 w} + \mathcal{O}\left(\frac{1}{\ln^5 w}\right), \quad w = \frac{s}{\Lambda^2}, \quad s \to \infty, \tag{6.9}$$

where $a_s^{(2)}(Q^2) = \alpha_s^{(2)}(Q^2)\beta_0/(4\pi)$ stands for the two-loop perturbative couplant (2.58) and $B_1 = \beta_1/\beta_0^2$, see Section 2.3 and Appendix A. Eqs. (6.8), (6.9) imply that, similar to the previous case, in the ultraviolet asymptotic $s \to \infty$ the difference between the expressions $R^{(2)}(s)$ (6.7) and $R_{\text{naive}}^{(2)}(s)$ (3.64) is beyond the accuracy of the two-loop level. Hence, for the energies s, which assume that the corrections of the order of $1/\ln^3(s/\Lambda^2)$ are negligible, the two-loop expression for the R-ratio (6.7) can be approximated by its naive form (3.64), too:

$$R^{(2)}(s) \simeq 1 + \sum_{j=1}^{2} d_j \left[a_s^{(2)}(|s|)\right]^j + \mathcal{O}\left(\frac{1}{\ln^3 w}\right) = R_{\text{naive}}^{(2)}(s) + \mathcal{O}\left(\frac{1}{\ln^3 w}\right). \tag{6.10}$$

Similar to the one-loop level discussed earlier the two-loop timelike effective expansion functions $A_{\text{TL},1}^{(2)}(s)$ and $A_{\text{TL},2}^{(2)}(s)$ are free of the unphysical singularities, whereas their reexpansions (6.8), (6.9) diverge at low energies. As before the timelike expansion functions $A_{\text{TL},1}^{(2)}(s)$ and $A_{\text{TL},2}^{(2)}(s)$ can be approximated by the sum of their naive forms and the respective π^2-terms only for the energies $\sqrt{s} > \exp(\pi/2)\Lambda$. The convergence of such sum is illustrated in Fig. 6.2, which displays the two-loop timelike expansion function of the second-order $A_{\text{TL},2}^{(2)}(s)$ (Eq. 5.39, solid curve) and its approximations (dashed curves) corresponding to various orders of the reexpansion (6.9). Specifically, the curve labeled "1" corresponds to the naive approximation $[a_s^{(2)}(|s|)]^2$, whereas the curves labeled "2," "3," and "4" additionally retain the π^2-terms proportional to $\ln^{-4} w$, $\ln^{-5} w$, and $\ln^{-6} w$, respectively. As one can infer from Fig. 6.2, the reexpansion (6.9) converges rather slowly at low and moderate energies. Besides,

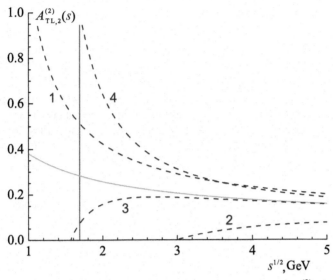

FIG. 6.2 Two-loop timelike effective expansion function of the second-order $A_{TL,2}^{(2)}(s)$ (Eq. 5.39, *solid curve*) and its approximations corresponding to various orders of the reexpansion (Eq. 6.9, *dashed curves*). *Numerical labels* specify the number of terms retained on the right-hand side of Eq. (6.9). The plotted functions are scaled by the factor of 10. *Vertical solid line* marks the boundary of the range of convergence of the reexpansion (6.9) at $\sqrt{s} = \exp(\pi/2)\Lambda$. The values of parameters: $n_f = 3$ active flavors, $\Lambda = 350\,\text{MeV}$.

even at relatively high energies $A_{TL,2}^{(2)}(s)$ considerably differs from $[a_s^{(2)}(|s|)]^2$, which enters the two-loop perturbative expression for the *R*-ratio (6.10). For example, at $\sqrt{s} = 5\,\text{GeV}$ $[a_s^{(2)}(|s|)]^2$ exceeds $A_{TL,2}^{(2)}(s)$ by about 30%, see Fig. 6.2.

It has to be emphasized that at the higher loop levels ($\ell \geq 3$) the π^2-terms are no longer negligible even in the ultraviolet asymptotic $s \to \infty$. In particular, the *R*-ratio of electron-positron annihilation into hadrons (5.27) at the three-loop level takes the following form:

$$R^{(3)}(s) = 1 + d_1 A_{TL,1}^{(3)}(s) + d_2 A_{TL,2}^{(3)}(s) + d_3 A_{TL,3}^{(3)}(s), \qquad (6.11)$$

where the timelike effective expansion functions $A_{TL,j}^{(\ell)}(s)$ are given by Eq. (5.26). As earlier, at high energies these functions can be approximated by

$$A_{TL,1}^{(3)}(s) \simeq a_s^{(3)}(|s|) - \frac{\pi^2}{3}\frac{1}{\ln^3 w} + \pi^2 \frac{B_1}{\ln^4 w}\left[\ln(\ln w) - \frac{5}{6}\right] + \mathcal{O}\left(\frac{1}{\ln^5 w}\right), \qquad (6.12)$$

$$A_{TL,2}^{(3)}(s) \simeq \left[a_s^{(3)}(|s|)\right]^2 - \frac{\pi^2}{\ln^4 w} + \mathcal{O}\left(\frac{1}{\ln^5 w}\right), \qquad (6.13)$$

$$A_{TL,3}^{(3)}(s) \simeq \left[a_s^{(3)}(|s|)\right]^3 + \mathcal{O}\left(\frac{1}{\ln^5 w}\right), \qquad w = \frac{s}{\Lambda^2}, \quad s \to \infty, \qquad (6.14)$$

where $a_s^{(3)}(Q^2) = \alpha_s^{(3)}(Q^2)\beta_0/(4\pi)$ denotes the three-loop perturbative couplant (2.80) and $B_1 = \beta_1/\beta_0^2$, see Section 2.3 and Appendix A. Similar to the one-loop and two-loop levels, the majority of the π^2-terms, which comprise the difference between the function $R^{(3)}(s)$ (6.11) and its naive form $R_{\text{naive}}^{(3)}(s)$ (3.64) at high energies $s \to \infty$, are beyond the given accuracy. However, one of those π^2-terms, namely, the second term on the right-hand side of Eq. (6.12), appears to be of the same order of magnitude as those of the original perturbative power series, and, therefore, cannot be neglected in the reexpansion of Eq. (6.11):

$$R^{(3)}(s) \simeq 1 + \sum_{j=1}^{3} d_j \left[a_s^{(3)}(|s|) \right]^j - \frac{\pi^2}{3} \frac{d_1}{\ln^3 w} + \mathcal{O}\left(\frac{1}{\ln^4 w}\right), \quad s \to \infty. \quad (6.15)$$

It is necessary to outline that the R-ratio of electron-positron annihilation into hadrons (5.27) by construction accounts for all the effects due to continuation of the spacelike theoretical results into the timelike domain. At the same time if one is interested in the behavior of the function $R(s)$ in the ultraviolet asymptotic $s \to \infty$ only, then Eq. (5.27) can be approximated by the sum of its naive form (3.64) and the nonnegligible π^2-terms (if any) at the given loop level, likewise Eq. (6.15). It is interesting to note here that at an arbitrary loop level the nonnegligible π^2-terms can be incorporated into the naive approximation of the function $R(s)$, the respective expansion coefficients d_j being properly amended.

In particular, for the case of Eq. (6.15) the aforementioned rearrangement is quite straightforward. Specifically, since at $s \to \infty$

$$\frac{1}{\ln^3 w} \simeq \left[a_s^{(3)}(|s|) \right]^3 + \mathcal{O}\left(\frac{1}{\ln^4 w}\right), \quad (6.16)$$

the π^2-term on the right-hand side of Eq. (6.15) can be absorbed[4] into the third term of the naive approximation of the strong correction to the R-ratio. In turn, this alters the three-loop perturbative coefficient d_3 and eventually leads to the following expression for the reexpanded function $R^{(3)}(s)$ (6.15) in the ultraviolet asymptotic:

$$R^{(3)}(s) \simeq 1 + \sum_{j=1}^{3} r_j \left[a_s^{(3)}(|s|) \right]^j + \mathcal{O}\left(\frac{1}{\ln^4 w}\right) = R_{\text{pert}}^{(3)}(s) + \mathcal{O}\left(\frac{1}{\ln^4 w}\right),$$

$$(6.17)$$

where $R_{\text{pert}}^{(\ell)}(s)$ is given by Eq. (3.67) and the coefficients r_j read

$$r_j = d_j - \delta_j. \quad (6.18)$$

Obviously, the first two coefficients r_j ($j = 1, 2$) coincide with those of the original perturbative power series:

$$\delta_1 = 0, \quad (6.19)$$

$$\delta_2 = 0, \quad (6.20)$$

---

4. Within the accuracy of the three-loop level.

whereas the third one is amended to properly accommodate the three-loop π^2-term, specifically [107, 108, 151],

$$\delta_3 = \frac{\pi^2}{3} d_1 B_0^2 = \frac{\pi^2}{3} d_1. \tag{6.21}$$

The three-loop perturbative approximation of the R-ratio (6.17) can be used for the energies s, which assume that the corrections of the order of $1/\ln^4(s/\Lambda^2)$ can be safely neglected. Recall that the approximation of the function $R(s)$ in the form of Eq. (6.17) is inapplicable for the energies $\sqrt{s} < \exp(\pi/2)\Lambda \simeq 4.81\Lambda$, with Λ being the QCD scale parameter. It also has to be noted that within the accuracy of the three-loop level the expressions (6.15), (6.17) are indistinguishable and either of them can be employed in practical applications.

6.2 GENERAL FEATURES

The results presented in the previous section imply that the rearrangement of reexpansion of the R-ratio of electron-positron annihilation into hadrons (5.27) in the ultraviolet asymptotic $s \to \infty$ becomes more tangled at the higher loop levels ($\ell \geq 4$). Specifically, in addition to the increasing number of the nonnegligible π^2-terms one also has to properly account for all the terms coming from the substitution of the form of Eq. (6.16). In particular, one part of such terms eliminates[5] the higher-order π^2-terms, which are proportional to the integer powers of $\ln(\ln w)$, whereas the other part directly contributes to the corresponding coefficients δ_j. To illustrate this issue in detail it is worthwhile to examine the high-energy behavior of the function $R(s)$ (5.27) at the higher loop levels.

The R-ratio of electron-positron annihilation into hadrons (5.27) at the five-loop level ($\ell = 5$) reads

$$R^{(5)}(s) = 1 + \sum_{j=1}^{5} d_j A_{\mathrm{TL},j}^{(5)}(s). \tag{6.22}$$

In the ultraviolet asymptotic $s \to \infty$ the involved five-loop timelike effective expansion functions $A_{\mathrm{TL},j}^{(5)}(s)$ (5.26) can be approximated by

$$
\begin{aligned}
A_{\mathrm{TL},1}^{(5)}(s) \simeq\; & a_s^{(5)}(|s|) - \frac{\pi^2}{3}\frac{1}{\ln^3 w} + \frac{\pi^2}{3}\frac{B_1}{\ln^4 w}\left[3\ln(\ln w) - \frac{5}{2}\right] \\
& + \frac{\pi^2}{3}\frac{1}{\ln^5 w}\left[-6B_1^2\ln^2(\ln w) + 13B_1^2\ln(\ln w) + \frac{3}{2}B_1^2 - 6B_2\right] \\
& + \frac{\pi^4}{5}\frac{1}{\ln^5 w} + \mathcal{O}\left(\frac{1}{\ln^6 w}\right),
\end{aligned}
\tag{6.23}
$$

---

5. That, in fact, allows one to perform the replacement of the form of Eq. (6.16) in the subsequent orders.

$$A_{TL,2}^{(5)}(s) \simeq \left[a_s^{(5)}(|s|)\right]^2 - \frac{\pi^2}{3}\frac{3}{\ln^4 w} + \frac{\pi^2}{3}\frac{B_1}{\ln^5 w}[12\ln(\ln w) - 7] + \mathcal{O}\left(\frac{1}{\ln^6 w}\right),$$
(6.24)

$$A_{TL,3}^{(5)}(s) \simeq \left[a_s^{(5)}(|s|)\right]^3 - \frac{\pi^2}{3}\frac{6}{\ln^5 w} + \mathcal{O}\left(\frac{1}{\ln^6 w}\right),$$
(6.25)

$$A_{TL,4}^{(5)}(s) \simeq \left[a_s^{(5)}(|s|)\right]^4 + \mathcal{O}\left(\frac{1}{\ln^6 w}\right),$$
(6.26)

$$A_{TL,5}^{(5)}(s) \simeq \left[a_s^{(5)}(|s|)\right]^5 + \mathcal{O}\left(\frac{1}{\ln^7 w}\right),$$
(6.27)

where $w = s/\Lambda^2$, $a_s^{(5)}(Q^2) = \alpha_s^{(5)}(Q^2)\beta_0/(4\pi)$ stands for the five-loop perturbative couplant (2.82), $B_1 = \beta_1/\beta_0^2$, and $B_2 = \beta_2/\beta_0^3$, see Section 2.3 and Appendix A. In particular, as one can infer from these equations, the π^2-terms do not appear at the orders of $1/\ln w$ and $1/\ln^2 w$, that immediately yields Eqs. (6.19), (6.20). By virtue of Eqs. (6.23)–(6.27) within the accuracy of the five-loop level the function $R(s)$ (6.22) can be approximated by the sum of its naive form (3.64) and the respective nonnegligible π^2-terms, specifically,

$$R^{(5)}(s) \simeq 1 + \sum_{j=1}^{5} d_j \left[a_s^{(5)}(|s|)\right]^j + r_\pi^{(5)}(s),$$
(6.28)

where

$$
\begin{aligned}
r_\pi^{(5)}(s) = &-\frac{\delta_3}{\ln^3 w} + \frac{\pi^2}{3}\frac{1}{\ln^4 w}\left\{d_1 B_1\left[3\ln(\ln w) - \frac{5}{2}\right] - 3d_2\right\}\\
&+ \frac{\pi^2}{3}\frac{1}{\ln^5 w}\left\{d_1 B_1^2\left[-6\ln^2(\ln w) + 13\ln(\ln w) + \frac{3}{2}\right] - 6d_1 B_2\right.\\
&\left. + d_2 B_1[12\ln(\ln w) - 7] - 6d_3\right\} + \frac{\pi^4}{5}\frac{d_1}{\ln^5 w} + \mathcal{O}\left(\frac{1}{\ln^6 w}\right)
\end{aligned}
$$
(6.29)

and δ_3 is given by Eq. (6.21).

Similar to the aforementioned replacement (6.16), the first term on the right-hand side of Eq. (6.29) can be substituted by the third power of the naive continuation of the five-loop perturbative couplant into the timelike domain, but in this case one has to retain all the nonnegligible within given accuracy terms, which comprise the difference between $1/\ln^3(s/\Lambda^2)$ and $[a_s^{(5)}(|s|)]^3$, namely

$$
\begin{aligned}
\frac{1}{\ln^3 w} \simeq &\left[a_s^{(5)}(|s|)\right]^3 + 3B_1\frac{\ln(\ln w)}{\ln^4 w}\\
&- \frac{3}{\ln^5 w}\left[2B_1^2\ln^2(\ln w) - B_1^2\ln(\ln w) - B_1^2 + B_2\right] + \mathcal{O}\left(\frac{1}{\ln^6 w}\right).
\end{aligned}
$$
(6.30)

The second term on the right-hand side of this equation eliminates the first term in the square brackets in the first line of Eq. (6.29), which allows one to perform the replacement similar to that of Eq. (6.30) at the subsequent (i.e., fourth) order. The first term in the square brackets in the second line of Eq. (6.30) cancels the first term in the square brackets in the second line of Eq. (6.29), the second term in the square brackets in the second line of Eq. (6.30) alters the second term in the square brackets in the second line of Eq. (6.29), whereas the third and fourth terms in the square brackets in the second line of Eq. (6.30) directly contribute to the fifth-order coefficient δ_5 (6.35), specifically:

$$r_\pi^{(5)}(s) = -\delta_3 \left[a_s^{(5)}(|s|) \right]^3 - \frac{\delta_4}{\ln^4 w} + \frac{\pi^2}{3} \frac{1}{\ln^5 w} \left\{ d_1 \left[10B_1^2 \ln(\ln w) - \frac{3}{2}B_1^2 - 3B_2 \right] \right.$$
$$\left. + d_2 \left[12B_1 \ln(\ln w) - 7B_1 \right] - 6d_3 \right\} + \frac{\pi^4}{5} \frac{d_1}{\ln^5 w} + \mathcal{O}\left(\frac{1}{\ln^6 w} \right),$$

$$(6.31)$$

where the corresponding fourth-order coefficient δ_4 reads [107, 108, 151]

$$\delta_4 = \frac{\pi^2}{3} \left(\frac{5}{2}d_1 B_0 B_1 + 3d_2 B_0^2 \right) = \frac{\pi^2}{3} \left(\frac{5}{2}d_1 B_1 + 3d_2 \right). \qquad (6.32)$$

In turn, within the accuracy of the five-loop level the second term on the right-hand side of Eq. (6.31) can be replaced by

$$\frac{1}{\ln^4 w} \simeq \left[a_s^{(5)}(|s|) \right]^4 + 4B_1 \frac{\ln(\ln w)}{\ln^5 w} + \mathcal{O}\left(\frac{1}{\ln^6 w} \right). \qquad (6.33)$$

The second term on the right-hand side of this equation eliminates both terms proportional to $\ln(\ln w)$ in Eq. (6.31), that casts the latter to

$$r_\pi^{(5)}(s) = -\delta_3 \left[a_s^{(5)}(|s|) \right]^3 - \delta_4 \left[a_s^{(5)}(|s|) \right]^4 - \frac{\delta_5}{\ln^5 w} + \mathcal{O}\left(\frac{1}{\ln^6 w} \right), \qquad (6.34)$$

where the fifth-order coefficient δ_5 takes the form [108, 431]

$$\delta_5 = \frac{\pi^2}{3} \left[\frac{3}{2}d_1 \left(B_1^2 + 2B_2 \right) + 7d_2 B_1 + 6d_3 \right] - \frac{\pi^4}{5}d_1. \qquad (6.35)$$

At the given loop level the replacement of the third term on the right-hand side of Eq. (6.34) is trivial

$$\frac{1}{\ln^5 w} \simeq \left[a_s^{(5)}(|s|) \right]^5 + \mathcal{O}\left(\frac{1}{\ln^6 w} \right), \qquad (6.36)$$

which finally brings the five-loop expression for π^2-terms (6.29) to the form that can easily be incorporated into the sum on the right-hand side of Eq. (6.28), namely,

$$r_\pi^{(5)}(s) = -\delta_3 \left[a_s^{(5)}(|s|) \right]^3 - \delta_4 \left[a_s^{(5)}(|s|) \right]^4 - \delta_5 \left[a_s^{(5)}(|s|) \right]^5 + \mathcal{O}\left(\frac{1}{\ln^6 w} \right).$$

(6.37)

Thus at the five-loop level the reexpansion of the function $R(s)$ (5.27) at high energies can be reduced to

$$R^{(5)}(s) \simeq 1 + \sum_{j=1}^{5} r_j \left[a_s^{(5)}(|s|) \right]^j + \mathcal{O}\left(\frac{1}{\ln^6 w} \right), \quad s \to \infty,$$

(6.38)

where $a_s^{(5)}(Q^2)$ is the five-loop perturbative couplant (2.82), $r_j = d_j - \delta_j$, and the coefficients δ_j of the first five orders have been specified in Eqs. (6.19)–(6.21), (6.32), (6.35). Also note that within the accuracy of the loop level on hand the expression (6.38) is indistinguishable from Eq. (6.28) and, in fact, either of them can be used in practical applications.

It is worthwhile to mention here that the approximation of the R-ratio of electron-positron annihilation into hadrons (5.27) by the power series in the naive timelike couplant $a_s(|s|)$ likewise Eq. (6.38) can only be considered as a formal representation of Eq. (5.27) in the form, which resembles the original perturbative power series for the Adler function. In particular, whereas in the expression for the function $R(s)$ (5.27) the terms corresponding to various orders of perturbation theory are clearly kept separately from each other, in the reexpansion (6.38) such terms appear to be intermixed. Basically, this is caused by the approximation of the timelike effective expansion functions $A_{\mathrm{TL},j}^{(\ell)}(s)$ (5.26) in the form of Eqs. (6.23)–(6.27) and the subsequent replacement of the resulting nonnegligible π^2-terms proportional to the inverse powers of $\ln w$ by the respective powers of the naive timelike couplant likewise Eqs. (6.30), (6.33), (6.36).

It is of a particular interest to outline that the aforementioned reduction of the reexpansion of the R-ratio of electron-positron annihilation into hadrons (5.27) in the ultraviolet asymptotic $s \to \infty$ to the form of a formal perturbative power series in the naive continuation of the perturbative couplant into the timelike domain $a_s(|s|)$ can be performed at an arbitrary loop level (see discussion later), specifically,

$$R^{(\ell)}(s) \simeq R_{\mathrm{pert}}^{(\ell)}(s) + \mathcal{O}\left[\frac{1}{\ln^{\ell+1}(s/\Lambda^2)} \right], \quad s \to \infty$$

(6.39)

where

$$R_{\mathrm{pert}}^{(\ell)}(s) = 1 + \sum_{j=1}^{\ell} r_j \left[a_s^{(\ell)}(|s|) \right]^j, \quad r_j = d_j - \delta_j,$$

(6.40)

see also Eqs. (3.66), (3.67). The function $R_{\mathrm{pert}}(s)$ (6.40) constitutes the perturbative approximation of the R-ratio. It is worthwhile to mention that, contrary to the original expression for the function $R(s)$ (5.27), its approximation $R_{\mathrm{pert}}(s)$ (6.40) contains infrared unphysical singularities and, therefore, is applicable at high energies only. Additionally, as it has been shown earlier, the approximation of the R-ratio in the form of Eq. (6.40) converges rather slowly at moderate energies and becomes invalid for $\sqrt{s} < \exp(\pi/2)\Lambda \simeq 4.81\Lambda$, with Λ being the QCD scale parameter.

Also note that the explicit expression for the ℓ-loop spectral function $\rho_{\text{pert}}^{(\ell)}(\sigma)$ (4.64) derived in Section 4.2 and Appendix B enables one to obtain the function $R(s)$ (5.27), which contains no unphysical singularities and thoroughly embodies the effects due to continuation of the spacelike perturbative results into the timelike domain, or "resummates" the respective π^2-terms to all orders, see Section 5.2. In turn, the reexpansion of R-ratio (5.27) at high energies results in the perturbative approximation $R_{\text{pert}}(s)$ (6.40), which entails all the difficulties listed earlier. At the same time it is worth mentioning that the expression $R_{\text{pert}}(s)$ (6.40) can be obtained in an equivalent way, which employs an approximate form of the corresponding spectral function from the very beginning, see also paper [431].

Specifically, at high energies the ℓ-loop strong correction to the R-ratio of electron-positron annihilation into hadrons (5.25) can be represented as

$$r^{(\ell)}(s) = \int_s^\infty \rho^{(\ell)}(\sigma)\frac{d\sigma}{\sigma} = \int_{\ln w}^\infty \bar{\rho}^{(\ell)}(y)\,dy, \quad y = \ln\left(\frac{\sigma}{\Lambda^2}\right), \quad w = \frac{s}{\Lambda^2} \quad (6.41)$$

(as noted earlier, the massless limit is assumed). In Eq. (6.41) $\bar{\rho}^{(\ell)}(y) = \rho^{(\ell)}[\Lambda^2\exp(y)]$ denotes the corresponding spectral function (4.37)

$$\bar{\rho}^{(\ell)}(y) = \frac{1}{2\pi i}[\bar{d}^{(\ell)}(y - i\pi) - \bar{d}^{(\ell)}(y + i\pi)], \quad (6.42)$$

with $\bar{d}^{(\ell)}(y) = d^{(\ell)}[\Lambda^2\exp(y)]$ being the ℓ-loop strong correction to the Adler function. Applying to the latter the Taylor expansion

$$\bar{d}^{(\ell)}(y \pm i\pi) = \bar{d}^{(\ell)}(y) + \sum_{n=1}^\infty \frac{(\pm i\pi)^n}{n!}\frac{d^n}{dy^n}\bar{d}^{(\ell)}(y), \quad |y| > \pi, \quad (6.43)$$

one can approximate[6] the spectral function (6.42) by

$$\bar{\rho}^{(\ell)}(y) = -\frac{d}{dy}\bar{d}^{(\ell)}(y) - \sum_{n=1}^\infty \frac{(-1)^n\pi^{2n}}{(2n+1)!}\frac{d^{2n+1}}{dy^{2n+1}}\bar{d}^{(\ell)}(y), \quad |y| > \pi. \quad (6.44)$$

Therefore the strong correction to the R-ratio (6.41) acquires the following form:

$$r^{(\ell)}(s) = d^{(\ell)}(|s|) + \sum_{n=1}^\infty \frac{(-1)^n\pi^{2n}}{(2n+1)!}\frac{d^{2n}}{dy^{2n}}\bar{d}^{(\ell)}(y)\Bigg|_{y=\ln w}, \quad w = \frac{s}{\Lambda^2}, \quad \sqrt{s} > \Lambda\exp\left(\frac{\pi}{2}\right).$$

$$(6.45)$$

---

6. It has to be noted here that Eqs. (6.43), (6.44) are only valid for $|y| > \pi$, that eventually bounds the convergence range of the perturbative approximation of R-ratio (6.40) to $\sqrt{s} > \exp(\pi/2)\Lambda$.

In particular, this equation illustrates the fact that at an arbitrary loop level the function $R(s)$ (5.27), being reexpanded at $s \to \infty$, reproduces term-by-term the naive continuation of the Adler function into the timelike domain (3.64) and additionally produces an infinite number of π^2-terms, see also discussion of this issue in Section 4.2.

In turn, the perturbative expression for the ℓ-loop strong correction to the Adler function entering Eq. (6.45) reads (3.57)

$$\bar{d}^{(\ell)}(y) = \sum_{j=1}^{\ell} d_j \left[\bar{a}^{(\ell)}(y) \right]^j, \tag{6.46}$$

with $\bar{a}^{(\ell)}(y) = a_{\rm s}^{(\ell)}[\Lambda^2 \exp(y)]$ being the ℓ-loop perturbative couplant, see Section 3.2 and Chapter 2 for the details. Then the renormalization group equation for the QCD invariant charge (2.12)

$$\frac{d}{dy} \bar{a}^{(\ell)}(y) = -\sum_{j=0}^{\ell-1} B_j \left[\bar{a}^{(\ell)}(y) \right]^{j+2}, \quad B_j = \frac{\beta_j}{\beta_0^{j+1}} \tag{6.47}$$

allows one to easily calculate the nth derivative of the jth power of the ℓ-loop couplant, specifically,

$$\frac{d^n}{dy^n}[\bar{a}^{(\ell)}(y)]^j = (-1)^n \sum_{k_1=0}^{\ell-1} \sum_{k_2=0}^{\ell-1} \cdots \sum_{k_n=0}^{\ell-1} \prod_{t=0}^{n-1} \left(j+t+\overbrace{k_1+k_2+\cdots+k_t}^{t \text{ terms}} \right)$$

$$\times \prod_{p=1}^{n} B_{k_p} \left[\bar{a}^{(\ell)}(y) \right]^{j+n+\overbrace{k_1+k_2+\cdots+k_n}^{n \text{ terms}}}. \tag{6.48}$$

Gathering Eqs. (6.45), (6.46), (6.48), one eventually arrives at the following approximation of the ℓ-loop strong correction to the R-ratio of electron-positron annihilation into hadrons (6.41):

$$r^{(\ell)}(s) = \sum_{j=1}^{\ell} d_j \left[a_{\rm s}^{(\ell)}(|s|) \right]^j - \sum_{j=1}^{\ell} d_j \sum_{n=1}^{\infty} \frac{(-1)^{n+1}\pi^{2n}}{(2n+1)!} \sum_{k_1=0}^{\ell-1} \sum_{k_2=0}^{\ell-1} \cdots \sum_{k_{2n}=0}^{\ell-1} \prod_{p=1}^{2n} B_{k_p}$$

$$\times \prod_{t=0}^{2n-1} \left(j+t+\overbrace{k_1+k_2+\cdots+k_t}^{t \text{ terms}} \right) \left[a_{\rm s}^{(\ell)}(|s|) \right]^{j+2n+\overbrace{k_1+k_2+\cdots+k_{2n}}^{2n \text{ terms}}},$$

$$\sqrt{s} > \Lambda \exp\left(\frac{\pi}{2}\right). \tag{6.49}$$

It is worth noting here that Eq. (6.49) explicitly proves that at any given loop level the reexpansion of the strong correction to R-ratio (5.25) at high energies can be reduced to the form of power series in the naive continuation of the strong running coupling to the timelike domain $\alpha_s^{(\ell)}(|s|)$. Additionally, this equation makes it evident that the π^2-terms do not appear in the first and second orders of perturbation theory because the lowest power of $a_s^{(\ell)}(|s|)$ in the second line of Eq. (6.49) is three (that corresponds to $j = 1$, $n = 1$, and $k_i = 0$). Eq. (6.49) also clearly demonstrates that the reexpansion of the R-ratio at high energies mixes up the coefficients d_j corresponding to various orders of perturbation theory.

The coefficients δ_j (6.18), which alter the corresponding perturbative coefficients d_j in the formal power series (6.40), incorporate all the nonnegligible at a given loop level π^2-terms. As one might note the "ℓ-loop" coefficient δ_ℓ ($\ell \geq 3$) involves the pertinent perturbative coefficients of the first ($\ell - 2$) orders, specifically, d_j ($1 \leq j \leq \ell - 2$) and β_j ($0 \leq j \leq \ell - 3$). In turn, this implies that, as mentioned earlier, starting from the three-loop level the contributions coming from the different orders of QCD perturbation theory appear to be all mixed together in the approximate expression (6.40), whereas in Eq. (5.27) the perturbative contributions of various orders are kept separately from each other. For example, the sixth-order coefficient [108, 431]

$$
\delta_6 = \frac{\pi^2}{3}\left[\frac{7}{2}d_1\left(B_1 B_2 + B_3\right) + 4d_2\left(B_1^2 + 2B_2\right) + \frac{27}{2}d_3 B_1 + 10d_4\right]
$$
$$
- \frac{\pi^4}{5}\left(\frac{77}{12}d_1 B_1 + 5d_2\right) \tag{6.50}
$$

incorporates the perturbative coefficients d_j and β_j up to the four-loop level (which are all known), whereas the seventh-order coefficient [431]

$$
\delta_7 = \frac{\pi^2}{3}\left[4d_1\left(B_1 B_3 + \frac{1}{2}B_2^2 + B_4\right) + 9d_2\left(B_1 B_2 + B_3\right)\right.
$$
$$
\left. + \frac{15}{2}d_3\left(B_1^2 + 2B_2\right) + 22d_4 B_1 + 15d_5\right]
$$
$$
- \frac{\pi^4}{5}\left[\frac{5}{6}d_1\left(17B_1^2 + 12B_2\right) + \frac{57}{2}d_2 B_1 + 15d_3\right] + \frac{\pi^6}{7}d_1 \tag{6.51}
$$

involves, in addition, the five-loop perturbative coefficients d_5 and β_4. As discussed earlier, the latter has been calculated in the paper [160], whereas for the former only numerical estimations are currently available, see also Sections 3.2 and 6.3. The explicit expressions for the coefficients δ_j up to the twelfth order ($1 \leq j \leq 12$) are gathered in Appendix C.

The numerical values of the coefficients δ_j (6.18) up to the six-loop level ($1 \leq j \leq 6$) are presented in Table 6.1. As one can infer from the comparison of the latter with the numerical values of the corresponding coefficients d_j of the Adler function perturbative expansion listed in Table 3.2, the coefficients δ_j can in no way be regarded as small corrections to d_j for $j \geq 3$. On the contrary, it appears that the values of coefficients δ_j (6.18) significantly exceed the values of respective perturbative coefficients d_j. As one can infer from Tables 3.2 and 6.2 (for higher-loop estimations see Tables 6.3 and 6.4 below), this fact eventually leads to an essential distortion of the original perturbative power series.

TABLE 6.1 Numerical Values of the Coefficients δ_j (6.18), Which Embody the Contributions of the Corresponding π^2-Terms, See Eqs. (6.19)–(6.21), (6.32), (6.35), (6.50), and Appendix C

n_f	δ_1	δ_2	δ_3	δ_4	$\delta_5 \times 10^{-1}$	$\delta_6 \times 10^{-1}$
0	0.0000	0.0000	1.1963	5.1127	2.0455	6.9081
1	0.0000	0.0000	1.2735	5.4298	1.8880	5.6819
2	0.0000	0.0000	1.3613	5.7583	1.7118	4.8532
3	0.0000	0.0000	1.4622	6.0851	1.3519	3.0365
4	0.0000	0.0000	1.5791	6.3850	0.6910	−0.3843
5	0.0000	0.0000	1.7165	6.6090	−0.3187	−4.5692
6	0.0000	0.0000	1.8799	6.6638	−2.1168	−12.0010

Specifically, both perturbative expressions for the Adler function $D_{\text{pert}}(Q^2)$ (3.57) and the R-ratio of electron-positron annihilation into hadrons $R_{\text{pert}}(s)$ (6.40) have the form of power series in the QCD couplant and their terms corresponding to the one-loop and two-loop levels are identical to each other. However, the three-loop and four-loop terms of $R_{\text{pert}}(s)$ (6.40) turn out to be substantially amplified and even sign-reversed with respect to the pertinent terms of $D_{\text{pert}}(Q^2)$ (3.57). Additionally, beyond the two-loop level (i.e., for $j \geq 3$) the coefficients δ_j increase rather rapidly as the loop level j increases. In particular, for all admissible values of active flavors n_f the three-loop coefficient δ_3 assumes values in the interval $1.2 \lesssim \delta_3 \lesssim 1.9$, whereas at the higher-loop levels the values of δ_j appear to be bound to much larger intervals, namely, $5.1 \lesssim \delta_4 \lesssim 6.7$ at the four-loop level, $-21 \lesssim \delta_5 \lesssim 20$ at the five-loop level, and $-120 \lesssim \delta_6 \lesssim 69$ at the six-loop level. In turn, this implies that the convergence of the R-ratio approximated by the power series (6.40) is worse than those of the function $R(s)$ given by Eq. (5.27) and of the perturbative approximation of the Adler function (3.57).

For example, for $n_f = 5$ active flavors at the four-loop level ($\ell = 4$) the value of the coefficient of perturbative expansion of the Adler function is $d_4 \simeq 0.649$ (see Table 3.2), whereas the value of the respective coefficient δ_4 exceeds it by an order of magnitude, specifically, $\delta_4 \simeq 6.609$ (see Eq. 6.32 and Table 6.1), that eventually leads to a rather large value of the corresponding coefficient r_4 in the perturbative approximation of the R-ratio of electron-positron annihilation into hadrons (6.40), namely, $r_4 \simeq -5.960$ (see Eq. 6.18 and Table 6.2). As previously discussed the coefficient δ_4 accounts for the contributions of the π^2-terms coming from the one-loop and two-loop levels only. In turn, the π^2-terms coming from the three-loop and four-loop levels contribute to the coefficients δ_j

TABLE 6.2 Numerical Values of the Coefficients r_j (6.18) of the Perturbative Approximation of the R-Ratio of Electron-Positron Annihilation Into Hadrons (6.40) up to the Four-Loop Level ($1 \leq j \leq 4$)

n_f	$r_1 = d_1$	$r_2 = d_2$	$r_3 = d_3 - \delta_3$	$r_4 = d_4 - \delta_4$
0	0.3636	0.2626	−0.3191	−2.7383
1	0.3871	0.2803	−0.4788	−3.2413
2	0.4138	0.3005	−0.6476	−3.6116
3	0.4444	0.3239	−0.9028	−4.1703
4	0.4800	0.3513	−1.2923	−5.0409
5	0.5217	0.3836	−1.8186	−5.9601
6	0.5714	0.4225	−2.6630	−7.5590

for $j \geq 5$ and $j \geq 6$, respectively. As one can infer from Tables 6.1 and 6.2 for the value of the naive timelike couplant $a_s(|s|) \simeq 0.1$ the contribution of the π^2-terms absorbed by the sixth-order coefficient δ_6 (6.50) to the perturbative approximation of the R-ratio (6.40) exceeds the contribution of the π^2-terms incorporated by the coefficient of the preceding (i.e., fifth) order δ_5 (6.35) by the factor of about 1.43, whereas their combined contribution to Eq. (6.40) alters its four-loop term by about 13%.

6.3   HIGHER-ORDER ESTIMATIONS

As mentioned in Chapters 2 and 3 there are a number of methods that enable one to estimate the numerical values of the uncalculated yet higher-order perturbative coefficients. Certainly, such methods, being rather approximate, can in no way be regarded as an alternative to the scrupulous calculation of the pertinent Feynman diagrams contributing to the process on hand. At the same time these methods may be somewhat useful in elucidating general features of the perturbative power series at the higher-loop levels. Thus let us explore the R-ratio of electron-positron annihilation into hadrons beyond currently available four-loop level.

The five-loop coefficient r_5 of the perturbative approximation of the R-ratio (6.40) is composed of two terms (6.18). Specifically, it consists of the uncalculated yet coefficient of the Adler function perturbative expansion d_5 and the coefficient δ_5 (6.35), which embodies the contributions due to the corresponding π^2-terms and involves known perturbative coefficients d_j ($1 \leq j \leq 3$) and β_j ($0 \leq j \leq 2$). As discussed in Section 3.2, the estimation

TABLE 6.3 Numerical Estimation of the Five-Loop Coefficients $\Pi_{5,1}$ (3.56) and d_5 (3.57) of the Adler Function Perturbative Expansion (See Paper [108]) and the resulting values of the five-loop coefficient r_5 (6.18) and the seven-loop coefficient δ_7 (6.51)

n_f	$\Pi_{5,1}$	d_5	$r_5 = d_5 - \delta_5$	δ_7
1	540	4.70	−14.2	7.02
2	308	3.74	−13.4	−35.7
3	145	2.52	−11.0	−82.5
4	45.7	1.16	−5.75	−115.7
5	0.661	0.0256	3.21	−83.0
6	4.38	0.267	21.4	142.5

of the five-loop coefficient d_5 was performed in paper [108] by making use of the effective charges method [283–286]. It is worth noting that the estimation of d_5 [108] assumes that its singlet part is negligible with respect to the nonsinglet one. In particular, this assumption is supported by the fact that at the three-loop ($j = 3$) and four-loop ($j = 4$) levels the singlet part amounts to several percents of the coefficient d_j for all admissible values[7] of active flavors n_f, except for the case of $j = 4$ and $n_f = 6$, see Table 3.3 for the details.

The numerical values of the estimated coefficient d_5 [108] and the resulting[8] coefficient r_5 (6.18) are given in Table 6.3. As one can infer from this table the aforementioned dominance of the π^2-terms gathered in the coefficients δ_j over the corresponding perturbative coefficients d_j persists beyond the four-loop level. This issue is also illustrated in Table 6.4, which presents the relative weight $(1 + |d_j/\delta_j|)^{-1} \times 100\%$ of the relevant π^2-terms in the coefficients r_j (6.18) of the R-ratio perturbative approximation (6.40) at various loop levels ($1 \leq j \leq 5$). Specifically, Table 6.4 clearly demonstrates that beyond the two-loop level (i.e., for $j \geq 3$) the major part of the coefficients r_j (6.18) is provided by the respective π^2-terms rather than by the Adler function perturbative expansion coefficients d_j. Additionally, as one can infer from Table 6.4, the dominance of the corresponding π^2-terms embodied in the coefficients δ_j over the perturbative coefficients d_j becomes more pronounced as

---

7. Recall that for $n_f = 3$ active flavors the singlet part of the perturbative coefficient d_j ($j \geq 3$) vanishes identically.

8. The numerical values of the coefficient δ_5 are listed in Table 6.1.

TABLE 6.4 The Relative Weight $(1 + |d_j/\delta_j|)^{-1} \times 100\%$ of the π^2-Terms in the Coefficients r_j (6.18) of the R-Ratio Perturbative Approximation (6.40) at Various Loop Levels $(1 \le j \le 5)$

n_f	$j = 1$	$j = 2$	$j = 3$	$j = 4$	$j = 5$ (Est.)
1	0.00%	0.00%	61.6%	71.3%	80.1%
2	0.00%	0.00%	65.6%	72.8%	82.1%
3	0.00%	0.00%	72.3%	76.1%	84.3%
4	0.00%	0.00%	84.6%	82.6%	85.6%
5	0.00%	0.00%	94.4%	91.1%	99.2%
6	0.00%	0.00%	70.6%	88.2%	98.8%

Notes: The last column ($j = 5$) involves the estimation of the five-loop coefficient d_5 performed in paper [108], see also Table 6.3.

the loop level j increases. In particular, at the three-loop level ($j = 3$) the relative weight of the coefficient δ_3 in r_3 is more than 60% for all admissible values of active flavors n_f, whereas at the four-loop ($j = 4$) and five-loop[9] ($j = 5$) levels it exceeds 70% and 80%, respectively.

Let us also address the seven-loop coefficient δ_7 (6.51). All its terms are known, except for the last one in the second line of Eq. (6.51), which involves the uncalculated yet five-loop coefficient d_5. As earlier for the latter the estimation performed in paper [108] can be used, that eventually results in the numerical values of the coefficient δ_7 (6.51) listed in Table 6.3. As one can infer from this table the seven-loop coefficient δ_7 may assume values in a rather large interval, namely, $-116 \lesssim \delta_7 \lesssim 143$, that substantially exceeds the values of the coefficients δ_j at the preceding loop levels, see Table 6.1. The results presented in Tables 6.3 and 6.4 also support the conclusions drawn earlier that the convergence of the R-ratio represented in the form of the power series (6.40) is worse than those of the original function $R(s)$ (5.27) and of the perturbative approximation of the Adler function (3.57).

To summarize all the issues discussed in this chapter, it is necessary to outline that one of the few, if not the only, advantage of using in practical applications the perturbative approximation of the R-ratio of electron-positron annihilation into hadrons (6.40) is that it is represented as a formal power series in the naive timelike couplant $a_s(|s|)$, which explicit form is available,

---

9. The values listed in the last column of Table 6.4 have been obtained by making use of the estimation of the five-loop perturbative coefficient d_5 performed in Ref. [108], see also Table 6.3.

see Chapter 2 and Appendix A. However, the approximation (6.40) contains infrared unphysical singularities, is applicable[10] only for the energies $\sqrt{s} > \exp(\pi/2)\Lambda \simeq 4.81\Lambda$, and slowly converges even at relatively high energies, which is primarily caused by a rapid growth of the higher-order coefficients δ_j that embody the corresponding π^2-terms. An apparent way to obviate all these difficulties is to abandon the perturbative approximation of the R-ratio in the form of Eq. (6.40) and to use instead the expression[11] (5.27), which contains no unphysical singularities, by construction accounts for all the effects[12] due to continuation of the spacelike perturbative results into the timelike domain (which appear to dominate over the corresponding perturbative contributions), and provides an enhanced stability of outcoming results with respect to the inclusion of the higher-loop corrections and to the choice of the subtraction scheme in the entire energy range.

---

10. Note that the perturbative approximation of the R-ratio (6.40), which constitutes the reexpansion of the expression $R(s)$ (5.27) in the ultraviolet asymptotic $s \to \infty$, converges rather slowly when the energy $\sqrt{s}$ approaches $\Lambda \exp(\pi/2)$.

11. The function $R(s)$ (5.27) can easily be evaluated in the way described in Appendix B.

12. It can be said that at a given loop level in the expression for the R-ratio (5.27) the respective π^2-terms are resummed to all orders.

Conclusions

The dispersion relations evidently play an essential role in a variety of issues of contemporary elementary particle physics. In particular, the dispersion relations not only enable one to study the strong interaction processes in the timelike domain in a self-consistent way, but also make it possible to get rid of some inherent difficulties of the perturbative approach to QCD and extend its range of applicability toward the infrared domain.

Appendix A

Perturbative QCD Running Coupling

As delineated in Chapter 2 the QCD invariant charge $\alpha_s(\mu^2) = g^2(\mu^2)/(4\pi)$ satisfies the renormalization group equation (2.1)

$$\frac{d\ln[g^2(\mu^2)]}{d\ln\mu^2} = \beta\left[g(\mu^2)\right] \tag{A.1}$$

and the perturbation theory still remains the basic method of calculation of the β function appearing on its right-hand side. At the same time one has to bear in mind that the perturbative approximation of the QCD β function (2.2)

$$\beta\left(g(\mu^2)\right) \simeq -\left\{\beta_0\left[\frac{g^2(\mu^2)}{16\pi^2}\right] + \beta_1\left[\frac{g^2(\mu^2)}{16\pi^2}\right]^2 + \beta_2\left[\frac{g^2(\mu^2)}{16\pi^2}\right]^3 + \cdots\right\} \tag{A.2}$$

is valid only for small values of the strong running coupling (that realizes in the ultraviolet domain) and hence becomes inapplicable at low energies. As mentioned in Chapter 2 for practical purposes it proves to be convenient to operate with the so-called QCD "couplant" $a(\mu^2) = \alpha(\mu^2)\beta_0/(4\pi)$. In this case Eqs. (A.1), (A.2) imply (2.12)

$$\frac{d\ln\left[a_s^{(\ell)}(\mu^2)\right]}{d\ln\mu^2} = -\sum_{j=0}^{\ell-1} B_j\left[a_s^{(\ell)}(\mu^2)\right]^{j+1}, \quad B_j = \frac{\beta_j}{\beta_0^{j+1}}. \tag{A.3}$$

At the one-loop level ($\ell = 1$) the renormalization group equation (A.3) can easily be integrated in finite limits, which eventually results in (2.19)

$$\alpha_s^{(1)}(Q^2) = \frac{4\pi}{\beta_0}\frac{1}{\ln z}, \quad z = \frac{Q^2}{\Lambda^2}. \tag{A.4}$$

In this equation Λ stands for the QCD scale parameter, which, as discussed in Section 2.2, absorbs all the terms independent of the energy scale Q^2 and of the value of the QCD invariant charge at Q^2. In turn, at the two-loop level ($\ell = 2$) the explicit solution to Eq. (A.3) can be represented as (2.55)

$$\alpha_s^{(2)}(Q^2) = -\frac{4\pi}{\beta_0}\frac{1}{B_1}\frac{1}{1 + W_{-1}\left\{-\exp\left[-\left(1 + B_1^{-1}\ln z\right)\right]\right\}}, \tag{A.5}$$

with $W_{-1}(x)$ being the respective branch of the multivalued Lambert W function (2.46). However, at the higher loop levels ($\ell \geq 3$) the solution to the renormalization group equation (A.3) cannot be expressed in terms of the presently known functions and can be found either numerically or by making use of approximate methods. For example, to make Eq. (A.3) explicitly solvable at the three-loop level, one may replace the perturbative β function on its right-hand side with the relevant Padé approximant, though it affects the infrared behavior of the resulting solution, see Section 2.3 for the details. At the same time in the majority of practical applications the iterative solution to Eq. (A.3) is commonly employed beyond the one-loop level (i.e., for $\ell \geq 2$).

In particular, the ℓ-loop iterative solution to the renormalization group equation for the QCD invariant charge (A.3) can be represented as the double sum (2.83)

$$\alpha_s^{(\ell)}(Q^2) = \frac{4\pi}{\beta_0}\sum_{n=1}^{\ell}\sum_{m=0}^{n-1} b_n^m \frac{\ln^m(\ln z)}{\ln^n z}, \quad z = \frac{Q^2}{\Lambda^2}, \tag{A.6}$$

where b_n^m (the integer superscript m is not to be confused with respective power) stands for the combination of the β function perturbative expansion coefficients (see later). Evidently, at an arbitrary loop level the iterative strong running coupling (A.6) possesses the infrared unphysical singularities in the kinematic variable Q^2. Note also that the terms, which constitute the difference between the exact solution to the renormalization group equation (A.3) and the iterative one (A.6), are beyond the accuracy of a given loop level, see Chapter 2 for the details.

It is worthwhile to mention that the ℓ-loop expression for the strong running coupling (A.6) includes all the coefficients b_n^m up to the ℓth order. At the first few loop levels $\alpha_s^{(\ell)}(Q^2)$ (A.6) acquires quite compact and simple form, whereas at the higher loop levels the expression (A.6) becomes rather cumbersome. The coefficients b_n^m up to the twelfth order ($1 \leq n \leq 12$, $0 \leq m \leq n-1$) are given in the following.

First-order coefficient:

$$b_1^0 = 1. \tag{A.7}$$

Second-order coefficients:

$$b_2^0 = 0, \tag{A.8}$$

$$b_2^1 = -B_1. \tag{A.9}$$

Third-order coefficients:

$$b_3^0 = -B_1^2 + B_2, \tag{A.10}$$

$$b_3^1 = -B_1^2, \tag{A.11}$$

$$b_3^2 = B_1^2. \tag{A.12}$$

Fourth-order coefficients:

$$b_4^0 = -\frac{1}{2}B_1^3 + \frac{1}{2}B_3, \tag{A.13}$$

$$b_4^1 = 2B_1^3 - 3B_1B_2, \tag{A.14}$$

$$b_4^2 = \frac{5}{2}B_1^3, \tag{A.15}$$

$$b_4^3 = -B_1^3. \tag{A.16}$$

Fifth-order coefficients:

$$b_5^0 = \frac{7}{6}B_1^4 - 3B_1^2B_2 - \frac{1}{6}B_1B_3 + \frac{5}{3}B_2^2 + \frac{1}{3}B_4, \tag{A.17}$$

$$b_5^1 = 4B_1^4 - 3B_1^2B_2 - 2B_1B_3, \tag{A.18}$$

$$b_5^2 = -\frac{3}{2}B_1^4 + 6B_2B_1^2, \tag{A.19}$$

$$b_5^3 = -\frac{13}{3}B_1^4, \tag{A.20}$$

$$b_5^4 = B_1^4. \tag{A.21}$$

Sixth-order coefficients:

$$b_6^0 = \frac{17}{12}B_1^5 - \frac{3}{2}B_1^3B_2 - \frac{23}{12}B_1^2B_3 - \frac{1}{12}B_1B_2^2 - \frac{1}{6}B_1B_4 + 2B_2B_3 + \frac{1}{4}B_5, \tag{A.22}$$

$$b_6^1 = -\frac{11}{6}B_1^5 + 12B_1^3B_2 - \frac{25}{3}B_1B_2^2 - \frac{7}{6}B_1^2B_3 - \frac{5}{3}B_1B_4, \tag{A.23}$$

$$b_6^2 = -\frac{23}{2}B_1^5 + \frac{27}{2}B_1^3B_2 + 5B_1^2B_3, \tag{A.24}$$

$$b_6^3 = -\frac{11}{6}B_1^5 - 10B_1^3B_2, \tag{A.25}$$

$$b_6^4 = \frac{77}{12}B_1^5, \tag{A.26}$$

$$b_6^5 = -B_1^5. \tag{A.27}$$

Seventh-order coefficients:

$$b_7^0 = -\frac{31}{30}B_1^6 + \frac{13}{2}B_1^4 B_2 - \frac{13}{60}B_1^3 B_3 - \frac{497}{60}B_1^2 B_2^2 - \frac{47}{30}B_1^2 B_4$$
$$- \frac{4}{5}B_1 B_2 B_3 - \frac{3}{20}B_1 B_5 + \frac{16}{5}B_2^3 + \frac{8}{5}B_2 B_4 + \frac{11}{20}B_3^2 + \frac{1}{5}B_6, \tag{A.28}$$

$$b_7^1 = -\frac{31}{3}B_1^6 + 21B_1^4 B_2 + \frac{31}{3}B_1^3 B_3 - \frac{47}{6}B_1^2 B_2^2 - \frac{2}{3}B_1^2 B_4 - 12B_1 B_2 B_3$$
$$- \frac{3}{2}B_1 B_5, \tag{A.29}$$

$$b_7^2 = -6B_1^6 - \frac{45}{2}B_1^4 B_2 + \frac{17}{2}B_1^3 B_3 + 25B_1^2 B_2^2 + 5B_1^2 B_4, \tag{A.30}$$

$$b_7^3 = \frac{127}{6}B_1^6 - 37B_1^4 B_2 - 10B_1^3 B_3, \tag{A.31}$$

$$b_7^4 = \frac{55}{6}B_1^6 + 15B_1^4 B_2, \tag{A.32}$$

$$b_7^5 = -\frac{87}{10}B_1^6, \tag{A.33}$$

$$b_7^6 = B_1^6. \tag{A.34}$$

Eighth-order coefficients:

$$b_8^0 = -\frac{349}{120}B_1^7 + \frac{29}{4}B_1^5 B_2 + \frac{317}{60}B_1^4 B_3 - \frac{37}{10}B_1^3 B_2^2 + \frac{1}{10}B_1^3 B_4$$
$$- \frac{231}{20}B_1^2 B_2 B_3 - \frac{7}{5}B_1^2 B_5 - \frac{23}{60}B_1 B_2^3 - \frac{49}{120}B_1 B_3^2 - \frac{9}{10}B_1 B_2 B_4$$
$$- \frac{2}{15}B_1 B_6 + \frac{19}{3}B_2^2 B_3 + \frac{17}{12}B_2 B_5 + \frac{5}{6}B_3 B_4 + \frac{1}{6}B_7, \tag{A.35}$$

$$b_8^1 = -\frac{31}{10}B_1^7 - \frac{49}{2}B_1^5 B_2 + \frac{237}{20}B_1^4 B_3 + \frac{1003}{20}B_1^3 B_2^2$$
$$+ \frac{103}{10}B_1^3 B_4 - \frac{32}{5}B_1^2 B_2 B_3 - \frac{9}{20}B_1^2 B_5 - \frac{112}{5}B_1 B_2^3$$
$$- \frac{77}{20}B_1 B_3^2 - \frac{56}{5}B_1 B_2 B_4 - \frac{7}{5}B_1 B_6, \tag{A.36}$$

$$b_8^2 = \frac{181}{6}B_1^7 - 96B_1^5B_2 - \frac{83}{3}B_1^4B_3 + \frac{629}{12}B_1^3B_2^2 + \frac{22}{3}B_1^3B_4$$
$$+ 42B_1^2B_2B_3 + \frac{21}{4}B_1^2B_5, \tag{A.37}$$

$$b_8^3 = \frac{211}{6}B_1^7 + \frac{31}{2}B_1^5B_2 - \frac{179}{6}B_1^4B_3 - \frac{175}{3}B_1^3B_2^2 - \frac{35}{3}B_1^3B_4, \tag{A.38}$$

$$b_8^4 = -\frac{223}{8}B_1^7 + \frac{319}{4}B_1^5B_2 + \frac{35}{2}B_1^4B_3, \tag{A.39}$$

$$b_8^5 = -\frac{323}{15}B_1^7 - 21B_1^5B_2, \tag{A.40}$$

$$b_8^6 = \frac{223}{20}B_1^7, \tag{A.41}$$

$$b_8^7 = -B_1^7. \tag{A.42}$$

Ninth-order coefficients:

$$b_9^0 = -\frac{17}{2520}B_1^8 - \frac{207}{20}B_1^6B_2 + \frac{4121}{1260}B_1^5B_3 + \frac{16577}{630}B_1^4B_2^2$$
$$+ \frac{3229}{630}B_1^4B_4 - \frac{97}{140}B_1^3B_2B_3 + \frac{8}{35}B_1^3B_5 - \frac{1861}{84}B_1^2B_2^3$$
$$- \frac{8957}{2520}B_1^2B_3^2 - \frac{137}{105}B_1^2B_6 - \frac{743}{70}B_1^2B_2B_4 - \frac{946}{315}B_1B_2^2B_3$$
$$- \frac{95}{126}B_1B_3B_4 - \frac{389}{420}B_1B_2B_5 - \frac{5}{42}B_1B_7 + \frac{2069}{315}B_2^4 + \frac{134}{35}B_2B_3^2$$
$$+ \frac{1787}{315}B_2^2B_4 + \frac{19}{63}B_4^2 + \frac{5}{7}B_3B_5 + \frac{46}{35}B_2B_6 + \frac{1}{7}B_8, \tag{A.43}$$

$$b_9^1 = \frac{121}{6}B_1^8 - \frac{165}{2}B_1^6B_2 - \frac{365}{12}B_1^5B_3 + \frac{319}{4}B_1^4B_2^2$$
$$+ \frac{19}{2}B_1^4B_4 + 86B_1^3B_2B_3 + \frac{43}{4}B_1^3B_5 - \frac{58}{3}B_1^2B_2^3$$
$$- \frac{7}{12}B_1^2B_3^2 - 4B_1^2B_2B_4 - \frac{1}{3}B_1^2B_6 - \frac{152}{3}B_1B_2^2B_3$$
$$- \frac{20}{3}B_1B_3B_4 - \frac{34}{3}B_1B_2B_5 - \frac{4}{3}B_1B_7, \tag{A.44}$$

$$b_9^2 = \frac{1277}{30}B_1^8 + 2B_1^6B_2 - \frac{1126}{15}B_1^5B_3 - \frac{8891}{60}B_1^4B_2^2$$
$$- \frac{508}{15}B_1^4B_4 + \frac{338}{5}B_1^3B_2B_3 + \frac{141}{20}B_1^3B_5 + \frac{448}{5}B_1^2B_2^3$$
$$+ \frac{77}{5}B_1^2B_3^2 + \frac{224}{5}B_1^2B_2B_4 + \frac{28}{5}B_1^2B_6, \tag{A.45}$$

$$b_9^3 = -\frac{815}{18}B_1^8 + \frac{791}{18}B_1^5B_3 + \frac{543}{2}B_1^6B_2 - \frac{1783}{9}B_1^4B_2^2$$
$$- \frac{281}{9}B_1^4B_4 - 14B_1^3B_5 - 112B_1^3B_2B_3, \tag{A.46}$$

$$b_9^4 = -\frac{2357}{24}B_1^8 + \frac{195}{4}B_1^6B_2 + \frac{463}{6}B_1^5B_3 + \frac{350}{3}B_1^4B_2^2 + \frac{70}{3}B_1^4B_4, \tag{A.47}$$

$$b_9^5 = \frac{346}{15}B_1^8 - \frac{743}{5}B_1^6B_2 - 28B_1^5B_3, \tag{A.48}$$

$$b_9^6 = \frac{1435}{36}B_1^8 + 28B_1^6B_2, \tag{A.49}$$

$$b_9^7 = -\frac{481}{35}B_1^8, \tag{A.50}$$

$$b_9^8 = B_1^8. \tag{A.51}$$

Tenth-order coefficients:

$$b_{10}^0 = \frac{12331}{2520}B_1^9 - \frac{889}{40}B_1^7B_2 - \frac{28739}{2520}B_1^6B_3 + \frac{16829}{630}B_1^5B_2^2$$
$$+ \frac{2741}{1260}B_1^5B_4 + \frac{6077}{140}B_1^4B_2B_3 + \frac{739}{140}B_1^4B_5 - \frac{872}{105}B_1^3B_2^3$$
$$+ \frac{353}{315}B_1^3B_3^2 + \frac{83}{70}B_1^3B_2B_4 + \frac{31}{105}B_1^3B_6 - \frac{122921}{2520}B_1^2B_2^2B_3$$
$$- \frac{1907}{315}B_1^2B_3B_4 - \frac{1121}{105}B_1^2B_2B_5 - \frac{209}{168}B_1^2B_7 - \frac{3293}{2520}B_1B_2^4$$
$$- \frac{881}{280}B_1B_2B_3^2 - \frac{4507}{1260}B_1B_2^2B_4 - \frac{163}{504}B_1B_4^2 - \frac{199}{280}B_1B_3B_5$$
$$- \frac{131}{140}B_1B_2B_6 - \frac{3}{28}B_1B_8 + \frac{92}{5}B_2^3B_3 + \frac{29}{40}B_3^3 + \frac{129}{20}B_2B_3B_4$$
$$+ \frac{45}{8}B_2^2B_5 + \frac{1}{2}B_4B_5 + \frac{13}{20}B_3B_6 + \frac{5}{4}B_2B_7 + \frac{1}{8}B_9, \tag{A.52}$$

$$b_{10}^1 = \frac{16991}{840}B_1^9 + \frac{213}{20}B_1^7B_2 - \frac{12569}{210}B_1^6B_3 - \frac{21989}{140}B_1^5B_2^2$$
$$- \frac{1282}{35}B_1^5B_4 + \frac{12913}{140}B_1^4B_2B_3 + \frac{1217}{140}B_1^4B_5 + \frac{15125}{84}B_1^3B_2^3$$
$$+ \frac{26381}{840}B_1^3B_3^2 + \frac{6407}{70}B_1^3B_2B_4 + \frac{1198}{105}B_1^3B_6 - \frac{2482}{105}B_1^2B_2^2B_3$$
$$+ \frac{5}{42}B_1^2B_3B_4 - \frac{1259}{420}B_1^2B_2B_5 - \frac{11}{42}B_1^2B_7 - \frac{2069}{35}B_1B_2^4$$

$$- \frac{1206}{35} B_1 B_2 B_3^2 - \frac{1787}{35} B_1 B_2^2 B_4 - \frac{19}{7} B_1 B_4^2 - \frac{45}{7} B_1 B_3 B_5$$

$$- \frac{414}{35} B_1 B_2 B_6 - \frac{9}{7} B_1 B_8, \tag{A.53}$$

$$b_{10}^2 = -\frac{2891}{60} B_1^9 + \frac{1493}{4} B_1^7 B_2 + \frac{7417}{120} B_1^6 B_3 - \frac{60847}{120} B_1^5 B_2^2$$

$$- \frac{4597}{60} B_1^5 B_4 - \frac{1597}{5} B_1^4 B_2 B_3 - \frac{1653}{40} B_1^4 B_5 + \frac{883}{5} B_1^3 B_2^3$$

$$+ \frac{721}{40} B_1^3 B_3^2 + \frac{314}{5} B_1^3 B_2 B_4 + \frac{71}{10} B_1^3 B_6 + 228 B_1^2 B_2^2 B_3$$

$$+ 30 B_1^2 B_3 B_4 + 51 B_1^2 B_2 B_5 + 6 B_1^2 B_7, \tag{A.54}$$

$$b_{10}^3 = -\frac{7784}{45} B_1^9 + \frac{531}{2} B_1^7 B_2 + \frac{24223}{90} B_1^6 B_3 + \frac{44359}{180} B_1^5 B_2^2$$

$$+ \frac{3167}{45} B_1^5 B_4 - \frac{1574}{5} B_1^4 B_2 B_3 - \frac{703}{20} B_1^4 B_5 - \frac{1344}{5} B_1^3 B_2^3$$

$$- \frac{231}{5} B_1^3 B_3^2 - \frac{84}{5} B_1^3 B_6 - \frac{672}{5} B_1^3 B_2 B_4, \tag{A.55}$$

$$b_{10}^4 = \frac{11}{3} B_1^9 - \frac{4497}{8} B_1^7 B_2 - \frac{521}{24} B_1^6 B_3 + \frac{6749}{12} B_1^5 B_2^2$$

$$+ \frac{1123}{12} B_1^5 B_4 + 252 B_1^4 B_2 B_3 + \frac{63}{2} B_1^4 B_5, \tag{A.56}$$

$$b_{10}^5 = \frac{23981}{120} B_1^9 - \frac{4727}{20} B_1^7 B_2 - \frac{1669}{10} B_1^6 B_3 - 210 B_1^5 B_2^2 - 42 B_1^5 B_4, \tag{A.57}$$

$$b_{10}^6 = \frac{947}{180} B_1^9 + \frac{2509}{10} B_1^7 B_2 + 42 B_1^6 B_3, \tag{A.58}$$

$$b_{10}^7 = -\frac{9099}{140} B_1^9 - 36 B_1^7 B_2, \tag{A.59}$$

$$b_{10}^8 = \frac{4609}{280} B_1^9, \tag{A.60}$$

$$b_{10}^9 = -B_1^9. \tag{A.61}$$

Eleventh-order coefficients:

$$b_{11}^0 = \frac{14887}{5040} B_1^{10} + \frac{7361}{840} B_1^8 B_2 - \frac{35909}{2520} B_1^7 B_3 - \frac{152629}{2520} B_1^6 B_2^2$$

$$- \frac{16549}{1260} B_1^6 B_4 + \frac{5113}{210} B_1^5 B_2 B_3 + \frac{1429}{840} B_1^5 B_5 + \frac{116911}{1260} B_1^4 B_2^3$$

$$+ \frac{25511}{1680} B_1^4 B_3^2 + \frac{4757}{105} B_1^4 B_2 B_4 + \frac{1753}{315} B_1^4 B_6 + \frac{11}{30} B_1^3 B_2^2 B_3$$

$$+ \frac{2743}{1260} B_1^3 B_2 B_5 + \frac{3727}{1260} B_1^3 B_3 B_4 + \frac{169}{504} B_1^3 B_7 - \frac{97817}{1680} B_1^2 B_2^4$$

$$- \frac{8961}{280} B_1^2 B_2 B_3^2 - \frac{728}{15} B_1^2 B_2^2 B_4 - \frac{4673}{420} B_1^2 B_2 B_6 - \frac{2441}{420} B_1^2 B_3 B_5$$

$$- \frac{407}{168} B_1^2 B_4^2 - \frac{101}{84} B_1^2 B_8 - \frac{3209}{315} B_1 B_2^3 B_3 - \frac{2651}{420} B_1 B_2 B_3 B_4$$

$$- \frac{337}{84} B_1 B_2^2 B_5 - \frac{79}{84} B_1 B_2 B_7 - \frac{38}{45} B_1 B_3^3 - \frac{41}{60} B_1 B_3 B_6 - \frac{7}{12} B_1 B_4 B_5$$

$$- \frac{7}{72} B_1 B_9 + \frac{883}{63} B_2^5 + \frac{249}{14} B_2^2 B_3^2 + \frac{1124}{63} B_2^3 B_4 + \frac{18}{7} B_2 B_4^2$$

$$+ \frac{43}{7} B_2 B_3 B_5 + \frac{122}{21} B_2^2 B_6 + \frac{76}{63} B_2 B_8 + \frac{7}{4} B_3^2 B_4 + \frac{11}{18} B_3 B_7$$

$$+ \frac{4}{9} B_4 B_6 + \frac{29}{144} B_5^2 + \frac{1}{9} B_{10}, \tag{A.62}$$

$$b_{11}^1 = -\frac{72337}{2520} B_1^{10} + \frac{2329}{10} B_1^8 B_2 + \frac{68281}{1260} B_1^7 B_3 - \frac{534481}{1260} B_1^6 B_2^2$$

$$- \frac{36781}{630} B_1^6 B_4 - \frac{47857}{140} B_1^5 B_2 B_3 - \frac{6173}{140} B_1^5 B_5 + \frac{7367}{28} B_1^4 B_2^3$$

$$+ \frac{5577}{70} B_1^4 B_2 B_4 + \frac{50903}{2520} B_1^4 B_3^2 + \frac{296}{35} B_1^4 B_6 + \frac{584821}{1260} B_1^3 B_2^2 B_3$$

$$+ \frac{14527}{140} B_1^3 B_2 B_5 + \frac{7643}{126} B_1^3 B_3 B_4 + \frac{341}{28} B_1^3 B_7 - \frac{58019}{1260} B_1^2 B_2^4$$

$$- \frac{419}{140} B_1^2 B_2 B_3^2 - \frac{9631}{630} B_1^2 B_2^2 B_4 - \frac{173}{70} B_1^2 B_2 B_6 + \frac{19}{28} B_1^2 B_3 B_5$$

$$+ \frac{131}{252} B_1^2 B_4^2 - \frac{3}{14} B_1^2 B_8 - 184 B_1 B_2^3 B_3 - \frac{129}{2} B_1 B_2 B_3 B_4 - \frac{225}{4} B_1 B_2^2 B_5$$

$$- \frac{25}{2} B_1 B_2 B_7 - \frac{29}{4} B_1 B_3^3 - \frac{13}{2} B_1 B_3 B_6 - 5 B_1 B_4 B_5 - \frac{5}{4} B_1 B_9, \tag{A.63}$$

$$b_{11}^2 = -\frac{125429}{840} B_1^{10} + 320 B_1^8 B_2 + \frac{303299}{840} B_1^7 B_3 + \frac{233741}{840} B_1^6 B_2^2$$

$$+ \frac{44741}{420} B_1^6 B_4 - \frac{109281}{140} B_1^5 B_2 B_3 - \frac{23741}{280} B_1^5 B_5 - \frac{303953}{420} B_1^4 B_2^3$$

$$- \frac{27639}{70} B_1^4 B_2 B_4 - \frac{29191}{210} B_1^4 B_3^2 - \frac{10489}{210} B_1^4 B_6 + \frac{7270}{21} B_1^3 B_2^2 B_3$$

$$+ \frac{5543}{84} B_1^3 B_2 B_5 + \frac{1235}{42} B_1^3 B_3 B_4 + \frac{307}{42} B_1^3 B_7 + \frac{2069}{7} B_1^2 B_2^4$$

$$+ \frac{1206}{7} B_1^2 B_2 B_3^2 + \frac{1787}{7} B_1^2 B_2^2 B_4 + \frac{414}{7} B_1^2 B_2 B_6 + \frac{225}{7} B_1^2 B_3 B_5$$
$$+ \frac{95}{7} B_1^2 B_4^2 + \frac{45}{7} B_1^2 B_8, \tag{A.64}$$

$$b_{11}^3 = -\frac{371}{30} B_1^{10} - \frac{2936}{3} B_1^8 B_2 + \frac{3787}{60} B_1^7 B_3 + \frac{58099}{30} B_1^6 B_2^2 + \frac{9773}{30} B_1^6 B_4$$
$$+ \frac{11248}{15} B_1^5 B_2 B_3 + \frac{513}{5} B_1^5 B_5 - \frac{12862}{15} B_1^4 B_2^3 - \frac{6377}{60} B_1^4 B_3^2$$
$$- \frac{5156}{15} B_1^4 B_2 B_4 - \frac{607}{15} B_1^4 B_6 - 760 B_1^3 B_2^2 B_3 - 100 B_1^3 B_3 B_4$$
$$- 170 B_1^3 B_2 B_5 - 20 B_1^3 B_7, \tag{A.65}$$

$$b_{11}^4 = \frac{3925}{9} B_1^{10} - \frac{9807}{8} B_1^8 B_2 - \frac{50009}{72} B_1^7 B_3 - \frac{3865}{72} B_1^6 B_2^2$$
$$- \frac{2965}{36} B_1^6 B_4 + 1039 B_1^5 B_2 B_3 + \frac{955}{8} B_1^5 B_5 + 672 B_1^4 B_2^3$$
$$+ \frac{231}{2} B_1^4 B_3^2 + 336 B_1^4 B_2 B_4 + 42 B_1^4 B_6, \tag{A.66}$$

$$b_{11}^5 = \frac{23101}{120} B_1^{10} + \frac{8879}{10} B_1^8 B_2 - \frac{7409}{60} B_1^7 B_3 - \frac{8009}{6} B_1^6 B_2^2$$
$$- \frac{1375}{6} B_1^6 B_4 - 504 B_1^5 B_2 B_3 - 63 B_1^5 B_5, \tag{A.67}$$

$$b_{11}^6 = -\frac{39337}{120} B_1^{10} + \frac{38689}{60} B_1^8 B_2 + \frac{1921}{6} B_1^7 B_3 + 350 B_1^6 B_2^2 + 70 B_1^6 B_4, \tag{A.68}$$

$$b_{11}^7 = -\frac{91361}{1260} B_1^{10} - \frac{2761}{7} B_1^8 B_2 - 60 B_1^7 B_3, \tag{A.69}$$

$$b_{11}^8 = \frac{54713}{560} B_1^{10} + 45 B_1^8 B_2, \tag{A.70}$$

$$b_{11}^9 = -\frac{4861}{252} B_1^{10}, \tag{A.71}$$

$$b_{11}^{10} = B_1^{10}. \tag{A.72}$$

Twelfth-order coefficients:

$$b_{12}^0 = -\frac{13373}{2016} B_1^{11} + \frac{1847}{35} B_1^9 B_2 + \frac{93043}{5040} B_1^8 B_3 - \frac{31011}{280} B_1^7 B_2^2$$

$$-\frac{3607}{280}B_1^7B_4 - \frac{1852}{15}B_1^6B_2B_3 - \frac{1622}{105}B_1^6B_5 + \frac{10727}{126}B_1^5B_2^3$$

$$+\frac{8413}{3360}B_1^5B_3^2 + \frac{257}{15}B_1^5B_2B_4 + \frac{1823}{1260}B_1^5B_6 + \frac{1169629}{5040}B_1^4B_2^2B_3$$

$$+\frac{10361}{360}B_1^4B_3B_4 + \frac{32047}{630}B_1^4B_2B_5 + \frac{373}{63}B_1^4B_7 - \frac{172793}{10080}B_1^3B_2^4$$

$$+\frac{1291}{105}B_1^3B_2B_3^2 + \frac{21031}{2520}B_1^3B_2^2B_4 + \frac{4093}{2520}B_1^3B_4^2 + \frac{809}{240}B_1^3B_3B_5$$

$$+\frac{304}{105}B_1^3B_2B_6 + \frac{38}{105}B_1^3B_8 - \frac{24771}{140}B_1^2B_2^3B_3 - \frac{8131}{1260}B_1^2B_3^3$$

$$-\frac{2067}{35}B_1^2B_2B_3B_4 - \frac{29801}{560}B_1^2B_2^2B_5 - \frac{533}{120}B_1^2B_4B_5 - \frac{7411}{1260}B_1^2B_3B_6$$

$$-\frac{9883}{840}B_1^2B_2B_7 - \frac{211}{180}B_1^2B_9 - \frac{49}{90}B_1B_4B_6 - \frac{839}{1260}B_1B_3B_7$$

$$-\frac{367}{1440}B_1B_5^2 - \frac{16195}{1008}B_1B_2^2B_3^2 - \frac{4}{45}B_1B_{10} - \frac{527}{42}B_1B_2^3B_4$$

$$-\frac{667}{168}B_1B_2^5 - \frac{1189}{1260}B_1B_2B_8 - \frac{2467}{840}B_1B_2B_4^2 - \frac{8209}{1260}B_1B_2B_3B_5$$

$$-\frac{1397}{315}B_1B_2^2B_6 - \frac{583}{252}B_1B_3^2B_4 + \frac{2011}{280}B_2B_3^3 + \frac{32237}{630}B_2^4B_3$$

$$+\frac{8161}{252}B_2^2B_3B_4 + \frac{283}{210}B_3B_4^2 + \frac{6929}{360}B_2^3B_5 + \frac{181}{112}B_3^2B_5 + \frac{169}{36}B_2B_4B_5$$

$$+\frac{173}{28}B_2B_3B_6 + \frac{1099}{180}B_2^2B_7 + \frac{41}{70}B_3B_8 + \frac{47}{40}B_2B_9 + \frac{7}{20}B_5B_6$$

$$+\frac{37}{90}B_4B_7 + \frac{1}{10}B_{11}, \tag{A.73}$$

$$b_{12}^1 = -\frac{308431}{5040}B_1^{11} + \frac{22933}{168}B_1^9B_2 + \frac{177187}{840}B_1^8B_3 + \frac{67773}{280}B_1^7B_2^2$$

$$+\frac{12053}{140}B_1^7B_4 - \frac{256057}{420}B_1^6B_2B_3 - \frac{52757}{840}B_1^6B_5 - \frac{68179}{90}B_1^5B_2^3$$

$$-\frac{740057}{5040}B_1^5B_3^2 - \frac{87923}{210}B_1^5B_2B_4 - \frac{16619}{315}B_1^5B_6 + \frac{579739}{1260}B_1^4B_2^2B_3$$

$$+\frac{3937}{140}B_1^4B_3B_4 + \frac{10057}{126}B_1^4B_2B_5 + \frac{4279}{504}B_1^4B_7 + \frac{599177}{1008}B_1^3B_2^4$$

$$+\frac{97733}{280}B_1^3B_2B_3^2 + \frac{65341}{126}B_1^3B_2^2B_4 + \frac{13693}{504}B_1^3B_4^2 + \frac{6784}{105}B_1^3B_3B_5$$

$$+\frac{1439}{12}B_1^3B_2B_6 + \frac{1093}{84}B_1^3B_8 - \frac{22661}{315}B_1^2B_2^3B_3 + \frac{367}{180}B_1^2B_3^3$$

$$+\frac{2071}{420}B_1^2B_2B_3B_4 - \frac{509}{42}B_1^2B_2^2B_5 + \frac{17}{12}B_1^2B_4B_5 + \frac{61}{60}B_1^2B_3B_6$$

$$- \frac{181}{84} B_1^2 B_2 B_7 - \frac{13}{72} B_1^2 B_9 - \frac{9713}{63} B_1 B_2^5 - \frac{2739}{14} B_1 B_2^2 B_3^2$$

$$- \frac{198}{7} B_1 B_2 B_4^2 - \frac{12364}{63} B_1 B_2^3 B_4 - \frac{77}{4} B_1 B_3^2 B_4 - \frac{319}{144} B_1 B_5^2$$

$$- \frac{473}{7} B_1 B_2 B_3 B_5 - \frac{1342}{21} B_1 B_2^2 B_6 - \frac{44}{9} B_1 B_4 B_6 - \frac{121}{18} B_1 B_3 B_7$$

$$- \frac{836}{63} B_1 B_2 B_8 - \frac{11}{9} B_1 B_{10}, \tag{A.74}$$

$$
\begin{aligned}
b_{12}^2 = {} & \frac{43133}{5040} B_1^{11} - \frac{19219}{20} B_1^9 B_2 + \frac{79403}{1260} B_1^8 B_3 + \frac{3290257}{1260} B_1^7 B_2^2 \\
& + \frac{269407}{630} B_1^7 B_4 + \frac{61573}{56} B_1^6 B_2 B_3 + \frac{22081}{140} B_1^6 B_5 - \frac{1823461}{840} B_1^5 B_2^3 \\
& - \frac{1260517}{5040} B_1^5 B_3^2 - \frac{23325}{28} B_1^5 B_2 B_4 - \frac{20257}{210} B_1^5 B_6 \\
& - \frac{5560631}{2520} B_1^4 B_2^2 B_3 - \frac{76663}{252} B_1^4 B_3 B_4 - \frac{423961}{840} B_1^4 B_2 B_5 \\
& - \frac{10025}{168} B_1^4 B_7 + \frac{1383049}{2520} B_1^3 B_2^4 + \frac{52849}{280} B_1^3 B_2 B_3^2 \\
& + \frac{427601}{1260} B_1^3 B_2^2 B_4 + \frac{5399}{504} B_1^3 B_3 B_4 + \frac{1591}{56} B_1^3 B_3 B_5 + \frac{10183}{140} B_1^3 B_2 B_6 \\
& + \frac{213}{28} B_1^3 B_8 + 1012 B_1^2 B_2^3 B_3 + \frac{319}{8} B_1^2 B_3^3 + \frac{1419}{4} B_1^2 B_2 B_3 B_4 \\
& + \frac{2475}{8} B_1^2 B_2^2 B_5 + \frac{55}{2} B_1^2 B_4 B_5 + \frac{143}{4} B_1^2 B_3 B_6 + \frac{275}{4} B_1^2 B_2 B_7 + \frac{55}{8} B_1^2 B_9, \\
& \tag{A.75}
\end{aligned}
$$

$$
\begin{aligned}
b_{12}^3 = {} & \frac{269711}{504} B_1^{11} - 2152 B_1^9 B_2 - \frac{635447}{504} B_1^8 B_3 + \frac{461833}{504} B_1^7 B_2^2 \\
& - \frac{16337}{252} B_1^7 B_4 + \frac{303407}{84} B_1^6 B_2 B_3 + \frac{69467}{168} B_1^6 B_5 + \frac{452615}{252} B_1^5 B_2^3 \\
& + \frac{101657}{252} B_1^5 B_3^2 + \frac{46369}{42} B_1^5 B_2 B_4 + \frac{17977}{126} B_1^5 B_6 - \frac{127850}{63} B_1^4 B_2^2 B_3 \\
& - \frac{26185}{126} B_1^4 B_3 B_4 - \frac{103813}{252} B_1^4 B_2 B_5 - \frac{5897}{126} B_1^4 B_7 - \frac{22759}{21} B_1^3 B_2^4 \\
& - \frac{4422}{7} B_1^3 B_2 B_3^2 - \frac{19657}{21} B_1^3 B_2^2 B_4 - \frac{1045}{21} B_1^3 B_4^2 - \frac{825}{7} B_1^3 B_3 B_5 \\
& - \frac{1518}{7} B_1^3 B_2 B_6 - \frac{165}{7} B_1^3 B_8, \tag{A.76}
\end{aligned}
$$

$$
\begin{aligned}
b_{12}^4 = {} & \frac{169243}{360} B_1^{11} + \frac{35171}{24} B_1^9 B_2 - \frac{625061}{720} B_1^8 B_3 - \frac{242074}{45} B_1^7 B_2^2 \\
& - \frac{352159}{360} B_1^7 B_4 - \frac{15347}{15} B_1^6 B_2 B_3 - \frac{6511}{40} B_1^6 B_5 + \frac{90901}{30} B_1^5 B_2^3
\end{aligned}
$$

$$+ \frac{97867}{240} B_1^5 B_3^2 + \frac{19219}{15} B_1^5 B_2 B_4 + \frac{9197}{60} B_1^5 B_6 + 2090 B_1^4 B_2^2 B_3$$

$$+ 275 B_1^4 B_3 B_4 + \frac{935}{2} B_1^4 B_2 B_5 + 55 B_1^4 B_7, \tag{A.77}$$

$$b_{12}^5 = -\frac{276097}{360} B_1^{11} + \frac{143393}{40} B_1^9 B_2 + \frac{101129}{72} B_1^8 B_3 - \frac{87605}{72} B_1^7 B_2^2$$

$$- \frac{1727}{36} B_1^7 B_4 - \frac{13949}{5} B_1^6 B_2 B_3 - \frac{2605}{8} B_1^6 B_5 - \frac{7392}{5} B_1^5 B_2^3$$

$$- \frac{2541}{10} B_1^5 B_3^2 - \frac{3696}{5} B_1^5 B_2 B_4 - \frac{462}{5} B_1^5 B_6, \tag{A.78}$$

$$b_{12}^6 = -\frac{490133}{720} B_1^{11} - 983 B_1^9 B_2 + \frac{196759}{360} B_1^8 B_3 + \frac{100699}{36} B_1^7 B_2^2$$

$$+ \frac{17645}{36} B_1^7 B_4 + 924 B_1^6 B_2 B_3 + \frac{231}{2} B_1^6 B_5, \tag{A.79}$$

$$b_{12}^7 = \frac{1115399}{2520} B_1^{11} - \frac{591239}{420} B_1^9 B_2 - \frac{23651}{42} B_1^8 B_3 - 550 B_1^7 B_2^2$$

$$- 110 B_1^7 B_4, \tag{A.80}$$

$$b_{12}^8 = \frac{397961}{2016} B_1^{11} + \frac{32891}{56} B_1^9 B_2 + \frac{165}{2} B_1^8 B_3, \tag{A.81}$$

$$b_{12}^9 = -\frac{233021}{1680} B_1^{11} - 55 B_1^9 B_2, \tag{A.82}$$

$$b_{12}^{10} = \frac{55991}{2520} B_1^{11}, \tag{A.83}$$

$$b_{12}^{11} = -B_1^{11}. \tag{A.84}$$

Appendix B

Perturbative Spectral Function

As outlined in Chapter 4 in the framework of dispersive approach to QCD the integral representations for the hadronic vacuum polarization function $\Pi(q^2)$, the R-ratio of electron-positron annihilation into hadrons, and the Adler function $D(Q^2)$ (Eqs. 4.34, 4.35, 4.36, respectively) involve the common spectral density $\rho(\sigma)$ (4.37). The perturbative part of the latter (4.55)

$$\rho_{\text{pert}}^{(\ell)}(\sigma) = \frac{1}{2\pi i} \lim_{\varepsilon \to 0_+} \left[d_{\text{pert}}^{(\ell)}(-\sigma - i\varepsilon) - d_{\text{pert}}^{(\ell)}(-\sigma + i\varepsilon) \right] \qquad (B.1)$$

can be calculated by making use of the perturbative expression for the strong correction to the Adler function (3.57)

$$d_{\text{pert}}^{(\ell)}(Q^2) = \sum_{j=1}^{\ell} d_j \left[a_{\text{s}}^{(\ell)}(Q^2) \right]^j. \qquad (B.2)$$

In this equation d_j stand for the corresponding coefficients of the Adler function perturbative expansion (see Section 3.2), $a_{\text{s}}^{(\ell)}(Q^2) = \alpha_{\text{s}}^{(\ell)}(Q^2)\beta_0/(4\pi)$ denotes the ℓ-loop perturbative QCD couplant (see Chapter 2 and Appendix A), $\beta_0 = 11 - 2n_f/3$ is the one-loop coefficient of the β function perturbative expansion (2.3), and n_f stands for the number of active flavors.

As discussed in Section 4.2 the perturbative spectral function $\rho_{\text{pert}}^{(\ell)}(\sigma)$ (B.1) can be explicitly calculated at any given loop level, the availability of the involved perturbative coefficients d_j and β_j being assumed. For this purpose it is worthwhile to represent Eq. (B.1) as

$$\rho_{\text{pert}}^{(\ell)}(\sigma) = \sum_{j=1}^{\ell} d_j \varrho_j^{(\ell)}(\sigma), \qquad (B.3)$$

where

$$\varrho_j^{(\ell)}(\sigma) = \frac{1}{2\pi i} \lim_{\varepsilon \to 0_+} \left\{ \left[a_{\text{s}}^{(\ell)}(-\sigma - i\varepsilon) \right]^j - \left[a_{\text{s}}^{(\ell)}(-\sigma + i\varepsilon) \right]^j \right\} \qquad (B.4)$$

stands for the contribution of the jth power of ℓ-loop perturbative couplant $a_{\text{s}}^{(\ell)}(Q^2)$ to Eq. (B.1), see Eqs. (4.64), (4.65). Then it is also convenient to define the real $a_{\text{Re}}^{(\ell)}(\sigma)$ and imaginary $a_{\text{Im}}^{(\ell)}(\sigma)$ parts of the ℓ-loop QCD couplant $a_s^{(\ell)}(Q^2)$ at the edges of its cut, specifically (4.66)

$$\lim_{\varepsilon \to 0_+} a_s^{(\ell)}(-\sigma \pm i\varepsilon) = a_{\text{Re}}^{(\ell)}(\sigma) \mp i\pi a_{\text{Im}}^{(\ell)}(\sigma), \tag{B.5}$$

with $a_{\text{Re}}^{(\ell)}(\sigma)$ and $a_{\text{Im}}^{(\ell)}(\sigma)$ being the real functions of their arguments. In turn, this enables one to simplify the ℓ-loop "partial" spectral function $\varrho_j^{(\ell)}(\sigma)$ (B.4) for any $j \geq 1$, namely (4.73)

$$\varrho_j^{(\ell)}(\sigma) = \sum_{k=0}^{K(j)} \binom{j}{2k+1} (-1)^k \pi^{2k} \left[a_{\text{Re}}^{(\ell)}(\sigma) \right]^{j-2k-1} \left[a_{\text{Im}}^{(\ell)}(\sigma) \right]^{2k+1}, \tag{B.6}$$

where

$$\binom{n}{m} = \frac{n!}{m!(n-m)!} \tag{B.7}$$

denotes the binomial coefficient,

$$K(j) = \frac{j}{2} + \frac{j \bmod 2}{2} - 1, \tag{B.8}$$

and $(j \bmod n)$ is the remainder on division of j by n, that is, $(j \bmod 2) = 0$ for even values of j and $(j \bmod 2) = 1$ for odd values of j.

By virtue of Eq. (2.83) the functions $a_{\text{Re}}^{(\ell)}(\sigma)$ and $a_{\text{Im}}^{(\ell)}(\sigma)$ entering Eq. (B.6) can be represented in the following form:

$$a_{\text{Re}}^{(\ell)}(\sigma) = \sum_{n=1}^{\ell} \sum_{m=0}^{n-1} b_n^m u_n^m(\sigma) \tag{B.9}$$

and

$$a_{\text{Im}}^{(\ell)}(\sigma) = \sum_{n=1}^{\ell} \sum_{m=0}^{n-1} b_n^m v_n^m(\sigma), \tag{B.10}$$

see Eqs. (4.80), (4.81). In these equations the numerical coefficients b_n^m constitute the combinations of the β function perturbative expansion coefficients (in particular, $b_1^0 = 1$, $b_2^0 = 0$, $b_2^1 = -B_1 = -\beta_1/\beta_0^2$, etc.), see Chapter 2 and Appendix A, whereas the functions $u_n^m(\sigma)$ and $v_n^m(\sigma)$ (the integer superscripts m are not to be confused with respective powers) are defined in Eqs. (4.98), (4.99), namely,

$$u_n^m(\sigma) = \begin{cases} u_n^0(\sigma), & \text{if } m = 0, \\ u_n^0(\sigma)u_0^m(\sigma) - \pi^2 v_n^0(\sigma)v_0^m(\sigma), & \text{if } m \geq 1, \end{cases} \tag{B.11}$$

and

$$v_n^m(\sigma) = \begin{cases} v_n^0(\sigma), & \text{if } m = 0, \\ v_n^0(\sigma)u_0^m(\sigma) + u_n^0(\sigma)v_0^m(\sigma), & \text{if } m \geq 1, \end{cases} \tag{B.12}$$

with $n \geq 1$ being assumed. In Eqs. (B.11), (B.12)

$$u_n^0(\sigma) = \frac{1}{(y^2 + \pi^2)^n} \sum_{k=0}^{K(n+1)} \binom{n}{2k} (-1)^k \pi^{2k} y^{n-2k}, \tag{B.13}$$

$$v_n^0(\sigma) = \frac{1}{(y^2 + \pi^2)^n} \sum_{k=0}^{K(n)} \binom{n}{2k+1} (-1)^k \pi^{2k} y^{n-2k-1}, \tag{B.14}$$

$$u_0^m(\sigma) = \sum_{k=0}^{K(m+1)} \binom{m}{2k} (-1)^k \pi^{2k} [L_1(y)]^{m-2k} [L_2(y)]^{2k}, \tag{B.15}$$

$$v_0^m(\sigma) = \sum_{k=0}^{K(m)} \binom{m}{2k+1} (-1)^{k+1} \pi^{2k} [L_1(y)]^{m-2k-1} [L_2(y)]^{2k+1}, \tag{B.16}$$

where

$$L_1(y) = \ln\sqrt{y^2 + \pi^2}, \tag{B.17}$$

$$L_2(y) = \frac{1}{2} - \frac{1}{\pi} \arctan\left(\frac{y}{\pi}\right), \tag{B.18}$$

the function $K(j)$ is given by Eq. (B.8), $y = \ln(\sigma/\Lambda^2)$, and it is assumed that $\arctan(x)$ is a monotone nondecreasing function of its argument, namely $-\pi/2 \leq \arctan(x) \leq \pi/2$ for $-\infty < x < \infty$, see Section 4.2 for the details. The expressions (B.3)–(B.18) enable one to explicitly calculate the perturbative spectral function $\rho_{\text{pert}}^{(\ell)}(\sigma)$ (B.1) at an arbitrary loop level.

For example, the four-loop ($\ell = 4$) perturbative spectral function $\rho_{\text{pert}}^{(4)}(\sigma)$ (B.1) can be represented as the sum of the four-loop partial spectral functions $\varrho_j^{(4)}(\sigma)$ (B.4) of the first four orders ($1 \leq j \leq 4$), specifically (B.3)

$$\rho_{\text{pert}}^{(4)}(\sigma) = d_1 \varrho_1^{(4)}(\sigma) + d_2 \varrho_2^{(4)}(\sigma) + d_3 \varrho_3^{(4)}(\sigma) + d_4 \varrho_4^{(4)}(\sigma). \tag{B.19}$$

Then the partial ℓ-loop spectral function $\varrho_j^{(\ell)}(\sigma)$ (B.4) of an arbitrary order j can be expressed in terms of the real $a_{\text{Re}}^{(\ell)}(\sigma)$ and imaginary $a_{\text{Im}}^{(\ell)}(\sigma)$ parts of the ℓ-loop perturbative QCD couplant $a_s^{(\ell)}(Q^2)$ at the edges of its cut by making use of Eq. (B.6). In particular, the first four relations (B.6) read

$$\varrho_1^{(\ell)}(\sigma) = a_{\text{Im}}^{(\ell)}(\sigma), \tag{B.20}$$

$$\varrho_2^{(\ell)}(\sigma) = 2a_{\text{Im}}^{(\ell)}(\sigma)a_{\text{Re}}^{(\ell)}(\sigma), \tag{B.21}$$

$$\varrho_3^{(\ell)}(\sigma) = a_{\text{Im}}^{(\ell)}(\sigma) \left\{ 3\left[a_{\text{Re}}^{(\ell)}(\sigma)\right]^2 - \pi^2 \left[a_{\text{Im}}^{(\ell)}(\sigma)\right]^2 \right\}, \tag{B.22}$$

$$\varrho_4^{(\ell)}(\sigma) = 4a_{\text{Im}}^{(\ell)}(\sigma)a_{\text{Re}}^{(\ell)}(\sigma) \left\{ \left[a_{\text{Re}}^{(\ell)}(\sigma)\right]^2 - \pi^2 \left[a_{\text{Im}}^{(\ell)}(\sigma)\right]^2 \right\}. \tag{B.23}$$

And finally at any given loop level the functions $a_{Re}^{(\ell)}(\sigma)$ and $a_{Im}^{(\ell)}(\sigma)$ entering Eq. (B.6) can be explicitly calculated by making use of Eqs. (B.9), (B.10). For instance, the four-loop functions $a_{Re}^{(4)}(\sigma)$ and $a_{Im}^{(4)}(\sigma)$ are given in Eqs. (B.36), (B.37), respectively.

The complete set of explicit expressions for real $a_{Re}^{(\ell)}(\sigma)$ and imaginary $a_{Im}^{(\ell)}(\sigma)$ parts of the ℓ-loop ($1 \leq \ell \leq 4$) perturbative QCD couplant $a_s^{(\ell)}(Q^2)$ at the edges of its cut is given in Eqs. (B.24)–(B.45). This set enables one to easily assemble the perturbative spectral function $\rho_{pert}^{(\ell)}(\sigma)$ (B.1) at the first four loop levels[1] and employ it in various practical applications (see also Appendix III of the review [195] and papers [350–352]).

At the one-loop level the real and imaginary parts of the one-loop perturbative QCD couplant $a_s^{(1)}(Q^2)$ (2.19) at the edges of its cut (B.5) have a quite simple form, specifically,

$$a_{Im}^{(1)}(\sigma) = \frac{1}{y^2 + \pi^2}, \quad y = \ln\left(\frac{\sigma}{\Lambda^2}\right), \tag{B.24}$$

$$a_{Re}^{(1)}(\sigma) = \frac{y}{y^2 + \pi^2}, \tag{B.25}$$

see also Eqs. (4.61)–(4.63).

At the two-loop level the functions (B.5) corresponding to the two-loop perturbative QCD couplant $a_s^{(2)}(Q^2)$ (2.78) take the following form:

$$a_{Im}^{(2)}(\sigma) = \frac{1}{(y^2 + \pi^2)^2}\left[2yS_1^{(2)}(y) + (\pi^2 - y^2)S_2^{(2)}(y)\right], \tag{B.26}$$

$$a_{Re}^{(2)}(\sigma) = \frac{1}{(y^2 + \pi^2)^2}\left[(y^2 - \pi^2)S_1^{(2)}(y) + 2\pi^2 yS_2^{(2)}(y)\right], \tag{B.27}$$

where

$$S_1^{(2)}(y) = y - \frac{B_1}{2}\ln(y^2 + \pi^2), \tag{B.28}$$

$$S_2^{(2)}(y) = 1 - B_1\left[\frac{1}{2} - \frac{1}{\pi}\arctan\left(\frac{y}{\pi}\right)\right], \tag{B.29}$$

and $B_j = \beta_j/\beta_0^{j+1}$ is the ratio of coefficients of the β function perturbative expansion (see Section 2.1).

In turn, at the three-loop level the functions (B.5) corresponding to the three-loop perturbative QCD couplant $a_s^{(3)}(Q^2)$ (2.80) can be represented as

$$a_{Im}^{(3)}(\sigma) = \frac{1}{(y^2 + \pi^2)^3}\left[(3y^2 - \pi^2)S_3^{(3)}(y) - y(y^2 - 3\pi^2)S_4^{(3)}(y)\right], \tag{B.30}$$

---

1. Formally, the set of Eqs. (B.24)–(B.45) also enables one to explicitly calculate the partial spectral function $\varrho_j^{(\ell)}(\sigma)$ (B.6) corresponding to any integer power j of the perturbative QCD couplant $a_s^{(\ell)}(Q^2)$ at first four loop levels $1 \leq \ell \leq 4$.

$$a_{\text{Re}}^{(3)}(\sigma) = \frac{1}{(y^2 + \pi^2)^3} \left[y(y^2 - 3\pi^2)S_3^{(3)}(y) + \pi^2(3y^2 - \pi^2)S_4^{(3)}(y) \right], \quad \text{(B.31)}$$

where

$$S_1^{(3)}(y) = \tfrac{1}{2} \ln(y^2 + \pi^2), \quad \text{(B.32)}$$

$$S_2^{(3)}(y) = \tfrac{1}{2} - \tfrac{1}{\pi} \arctan\left(\tfrac{y}{\pi}\right), \quad \text{(B.33)}$$

$$S_3^{(3)}(y) = y^2 - \pi^2 - B_1 \left[y S_1^{(3)}(y) - \pi^2 S_2^{(3)}(y) \right]$$
$$+ B_1^2 \left\{ S_1^{(3)}(y) \left[S_1^{(3)}(y) - 1 \right] - \pi^2 \left[S_2^{(3)}(y) \right]^2 - 1 \right\} + B_2, \quad \text{(B.34)}$$

$$S_4^{(3)}(y) = 2y - B_1 \left[S_1^{(3)}(y) + y S_2^{(3)}(y) \right] + B_1^2 S_2^{(3)}(y) \left[2 S_1^{(3)}(y) - 1 \right]. \quad \text{(B.35)}$$

At the four-loop level the functions (B.5) corresponding to the four-loop perturbative QCD couplant $a_s^{(4)}(Q^2)$ (2.81) acquire the form

$$a_{\text{Im}}^{(4)}(\sigma) = \frac{1}{(y^2 + \pi^2)^4} \left\{ 4y(y^2 - \pi^2) \left[S_3^{(4)}(y) + S_5^{(4)}(y) + S_7^{(4)}(y) \right] \right.$$
$$\left. + \left[4\pi^2 y^2 - (y^2 - \pi^2)^2 \right] \left[S_4^{(4)}(y) + S_6^{(4)}(y) + S_8^{(4)}(y) \right] \right\}, \quad \text{(B.36)}$$

$$a_{\text{Re}}^{(4)}(\sigma) = \frac{1}{(y^2 + \pi^2)^4} \left\{ \left[(y^2 - \pi^2)^2 - 4\pi^2 y^2 \right] \left[S_3^{(4)}(y) + S_5^{(4)}(y) + S_7^{(4)}(y) \right] \right.$$
$$\left. + 4\pi^2 y(y^2 - \pi^2) \left[S_4^{(4)}(y) + S_6^{(4)}(y) + S_8^{(4)}(y) \right] \right\}, \quad \text{(B.37)}$$

where

$$S_1^{(4)}(y) = \frac{1}{2} \ln(y^2 + \pi^2), \quad \text{(B.38)}$$

$$S_2^{(4)}(y) = \frac{1}{2} - \frac{1}{\pi} \arctan\left(\frac{y}{\pi}\right), \quad \text{(B.39)}$$

$$S_3^{(4)}(y) = B_1^2 y \left\{ S_1^{(4)}(y) \left[S_1^{(4)}(y) - 1 \right] - \pi^2 \left[S_2^{(4)}(y) \right]^2 + \frac{B_2}{B_1^2} - 1 \right\}$$
$$- B_1^2 \pi^2 S_2^{(4)}(y) \left[2 S_1^{(4)}(y) - 1 \right], \quad \text{(B.40)}$$

$$S_4^{(4)}(y) = B_1^2 \left\{ S_1^{(4)}(y) \left[S_1^{(4)}(y) - 1 \right] - \pi^2 \left[S_2^{(4)}(y) \right]^2 \right.$$
$$\left. + y S_2^{(4)}(y) \left[2 S_1^{(4)}(y) - 1 \right] + \frac{B_2}{B_1^2} - 1 \right\}, \quad \text{(B.41)}$$

$$S_5^{(4)}(y) = B_1^3 \left\{ S_1^{(4)}(y) \left[3\pi^2 \left(S_2^{(4)}(y) \right)^2 - \left(S_1^{(4)}(y) \right)^2 \right] \right.$$
$$\left. + \frac{5}{2} \left[\left(S_1^{(4)}(y) \right)^2 - \pi^2 \left(S_2^{(4)}(y) \right)^2 \right] + S_1^{(4)}(y) \left(2 - 3\frac{B_2}{B_1^2} \right) + \frac{1}{2} \left(\frac{B_3}{B_1^3} - 1 \right) \right\}, \quad \text{(B.42)}$$

$$S_6^{(4)}(y) = B_1^3 S_2^{(4)}(y) \left\{ \pi^2 \left[S_2^{(4)}(y) \right]^2 - 3 \left[S_1^{(4)}(y) \right]^2 + 5 S_1^{(4)}(y) - 3 \frac{B_2}{B_1^2} + 2 \right\},$$

(B.43)

$$S_7^{(4)}(y) = y(y^2 - 3\pi^2) + B_1 \left[2\pi^2 y S_2^{(4)}(y) - (y^2 - \pi^2) S_1^{(4)}(y) \right],$$ (B.44)

$$S_8^{(4)}(y) = 3y^2 - \pi^2 - B_1 \left[(y^2 - \pi^2) S_2^{(4)}(y) + 2y S_1^{(4)}(y) \right].$$ (B.45)

Let us address now the computation of the strong corrections to the hadronic vacuum polarization function (4.34), the R-ratio of electron-positron annihilation into hadrons (4.35), and the Adler function (4.36). Basically, the complicacy of the described earlier expression for the perturbative spectral function $\rho_{\text{pert}}^{(\ell)}(\sigma)$ (B.1) implies that the integration in Eqs. (4.34)–(4.36) can only be performed by making use of the numerical methods. Nonetheless, the latter entails no technical difficulties since the strong corrections (4.34)–(4.36) constitute quite simple one-dimensional integrals of smooth finite functions of respective variables. Moreover, the ability to explicitly calculate the involved spectral function (B.1) at an arbitrary loop level and the development of the ready-to-use computational packages [350–352] substantially facilitate[2] the employment of the dispersively improved perturbation theory in a variety of practical applications.

First of all to compute the strong corrections $\Delta p(q^2, q_0^2)$ (4.34), $r(s)$ (4.35), and $d(Q^2)$ (4.36), it is obviously convenient to operate with the dimensionless variables

$$z = \frac{Q^2}{\Lambda^2}, \quad z_0 = \frac{Q_0^2}{\Lambda^2}, \quad w = \frac{s}{\Lambda^2}, \quad \chi = \frac{m^2}{\Lambda^2}.$$ (B.46)

Here $Q^2 = -q^2 > 0$ and $Q_0^2 = -q_0^2 > 0$ stand for the spacelike kinematic variables, $s = q^2 > 0$ denotes the timelike kinematic variable, m is the value of the hadronic production threshold, and Λ stands for the QCD scale parameter. It is also worthwhile to redefine the spectral function $\rho(\sigma)$ appearing in Eqs. (4.34)–(4.36) in the following way:

$$\rho_y(y) = \rho(\sigma)\big|_{\sigma=\Lambda^2 e^y}, \quad \rho_x(x) = \rho_y(y)\big|_{y=1/x}, \quad r(x) = \frac{1}{x^2} \rho_x(x).$$ (B.47)

Then the strong correction to the hadronic vacuum polarization function (4.34) can be represented in the following suitable for the numerical integration form:

---

2. One may also mention such attempts to simplify the calculations in the framework of APT [331, 353–355] as the tabulation of the values of the massless spacelike and timelike expansion functions for few fixed sets of input parameters [432], as well as the construction of interpolating functions, which approximate the latter within reasonable accuracy [433, 434].

$$\Delta p(q^2, q_0^2) = \int\limits_{m^2}^{\infty} \rho(\sigma) \ln\left(\frac{\sigma - q^2}{\sigma - q_0^2} \frac{m^2 - q_0^2}{m^2 - q^2}\right) \frac{d\sigma}{\sigma} \tag{B.48}$$

$$= \begin{cases} I_1^{HP}\left(-1, \frac{1}{\ln\chi}\right) + I_2^{HP}(-1, 1) + I_3^{HP}(0, 1), & \ln\chi < -1, \\ I_2^{HP}(\ln\chi, 1) + I_3^{HP}(0, 1), & -1 \le \ln\chi < 1, \\ I_3^{HP}\left(0, \frac{1}{\ln\chi}\right), & 1 \le \ln\chi, \end{cases} \tag{B.49}$$

where

$$I_1^{HP}(x_1, x_2) = \int\limits_{x_1}^{x_2} r(x) \ln\left(\frac{e^{1/x} + z}{e^{1/x} + z_0} \frac{\chi + z_0}{\chi + z}\right) dx, \tag{B.50}$$

$$I_2^{HP}(y_1, y_2) = \int\limits_{y_1}^{y_2} \rho(y) \ln\left(\frac{e^y + z}{e^y + z_0} \frac{\chi + z_0}{\chi + z}\right) dy, \tag{B.51}$$

$$I_3^{HP}(x_1, x_2) = \int\limits_{x_1}^{x_2} r(x) \ln\left(\frac{1 + ze^{-1/x}}{1 + z_0 e^{-1/x}} \frac{\chi + z_0}{\chi + z}\right) dx. \tag{B.52}$$

It is worth mentioning that the arguments of the logarithms in Eqs. (B.50), (B.52) are given in the form, which assures an accurate numerical assessment of their numerators and denominators for, respectively, small negative and small positive values of the integration variable x. In turn, the strong correction to the R-ratio of electron-positron annihilation into hadrons (4.35) can be rewritten as

$$r(s) = \theta(s - m^2) \int\limits_{s}^{\infty} \rho(\sigma) \frac{d\sigma}{\sigma} \tag{B.53}$$

$$= \begin{cases} I_1^{TL}\left(-1, \frac{1}{\ln w}\right) + I_2^{TL}(-1, 1) + I_1^{TL}(0, 1), & \ln\chi \le \ln w < -1, \\ I_2^{TL}(\ln w, 1) + I_1^{TL}(0, 1), & -1 \le \ln w < 1, \\ I_1^{TL}\left(0, \frac{1}{\ln w}\right), & 1 \le \ln w, \end{cases} \tag{B.54}$$

where

$$I_1^{TL}(x_1, x_2) = \theta(w - \chi) \int\limits_{x_1}^{x_2} r(x) dx, \tag{B.55}$$

$$I_2^{TL}(y_1, y_2) = \theta(w - \chi) \int\limits_{y_1}^{y_2} \rho_y(y) dy. \tag{B.56}$$

And finally the strong correction to the Adler function (4.36) acquires the following form:

$$d(Q^2) = \frac{Q^2}{Q^2 + m^2} \int\limits_{m^2}^{\infty} \rho(\sigma) \frac{\sigma - m^2}{\sigma + Q^2} \frac{d\sigma}{\sigma} \tag{B.57}$$

$$= \begin{cases} \frac{z}{z+\chi} \left[I_1^{SL}\left(-1, \frac{1}{\ln\chi}\right) + I_2^{SL}(-1, 1) + I_3^{SL}(0, 1) \right], & \ln\chi < -1, \\ \frac{z}{z+\chi} \left[I_2^{SL}(\ln\chi, 1) + I_3^{SL}(0, 1) \right], & -1 \le \ln\chi < 1, \\ \frac{z}{z+\chi} I_3^{SL}\left(0, \frac{1}{\ln\chi}\right), & 1 \le \ln\chi, \end{cases} \tag{B.58}$$

where

$$I_1^{SL}(x_1, x_2) = \int\limits_{x_1}^{x_2} r(x) \frac{e^{1/x} - \chi}{e^{1/x} + z} dx, \tag{B.59}$$

$$I_2^{SL}(y_1, y_2) = \int\limits_{y_1}^{y_2} \rho_y(y) \frac{e^y - \chi}{e^y + z} dy, \tag{B.60}$$

$$I_3^{SL}(x_1, x_2) = \int\limits_{x_1}^{x_2} r(x) \frac{1 - \chi e^{-1/x}}{1 + z e^{-1/x}} dx. \tag{B.61}$$

As earlier, the integrands in Eqs. (B.59), (B.61) are written in the form, which secures an accurate numerical assessment of their numerators and denominators for, respectively, small negative and small positive values of x.

The numerical integration in the equations presented earlier can easily be performed by making use of, for example, an adaptive Gaussian quadrature method. The detailed description of computation of the strong corrections $\Delta p(q^2, q_0^2)$ (B.48), $r(s)$ (B.53), and $d(Q^2)$ (B.57) can be found in papers [350–352]. It is also worthwhile to mention that similar subjects have been addressed in papers [435–438].

Appendix C

Perturbative Approximation of *R*-Ratio

As discussed in Chapters 4 and 5 since the perturbation theory is not directly applicable to the study of QCD observables depending on the timelike kinematic variable, the only consistent way to calculate the R-ratio of electron-positron annihilation into hadrons is to employ the relevant dispersion relations (4.1)–(4.6). The proper continuation of the spacelike theoretical results into the timelike domain leads to an essential distortion of the original perturbative power series, specifically (4.35),

$$R(s) = R^{(0)}(s) + \theta(s - m^2) \int\limits_{s}^{\infty} \rho(\sigma) \frac{d\sigma}{\sigma}. \tag{C.1}$$

In this equation $s = q^2 \geq 0$ is the timelike kinematic variable (namely, the center-of-mass energy squared), the leading-order term is given by Eq. (4.40), $\theta(x)$ stands for the Heaviside unit step function (1.27), m denotes the total mass of the lightest allowed hadronic final state, and $\rho(\sigma)$ is the spectral density (4.37). The perturbative part of the latter can be expressed in terms of the strong correction to the Adler function (3.57), namely (4.55),

$$\rho_{\text{pert}}^{(\ell)}(\sigma) = \frac{1}{2\pi i} \lim_{\varepsilon \to 0_+} \left[d_{\text{pert}}^{(\ell)}(-\sigma - i\varepsilon) - d_{\text{pert}}^{(\ell)}(-\sigma + i\varepsilon) \right]. \tag{C.2}$$

The relations derived in Section 4.2 and summarized in Appendix B enable one to explicitly calculate the perturbative spectral function (C.2) at an arbitrary loop level. It appears that the effects due to continuation of the spacelike perturbative results into the timelike domain make the resulting function $R(s)$ (C.1) completely different from its naive form (3.64). Moreover, these effects override the corresponding perturbative contributions even in the ultraviolet asymptotic $s \to \infty$ and thereby play a key role in the studies of R-ratio of electron-positron annihilation into hadrons and the related quantities in the entire energy range, see Section 5.2 and Chapter 6 for the details.

As delineated in Chapters 4 and 5 the expression for the R-ratio (C.1) contains no unphysical singularities, by construction incorporates all the effects due to continuation of the spacelike theoretical results into the timelike domain (or resummates the corresponding π^2-terms to all orders), and provides an enhanced stability of $R(s)$ with respect to the higher-loop corrections and to the choice of the subtraction scheme. As for the integration in Eq. (C.1), it can easily be performed either explicitly or numerically by making use of methods and computational tools discussed in Appendix B. At the same time as noted in Chapter 6 the reexpansion of function $R(s)$ (C.1) in the ultraviolet asymptotic $s \to \infty$ can be reduced to the form of power series in the naive timelike couplant at a given loop level, specifically (6.40)

$$R_{\text{pert}}^{(\ell)}(s) = 1 + \sum_{j=1}^{\ell} r_j \left[a_s^{(\ell)}(|s|) \right]^j, \quad r_j = d_j - \delta_j. \tag{C.3}$$

In this equation $a_s^{(\ell)}(Q^2)$ denotes the ℓ-loop QCD couplant (see Chapter 2 and Appendix A), d_j denote the Adler function perturbative expansion coefficients (3.57), and the coefficients δ_j account for the corresponding π^2-terms. However, one has to be aware that the perturbative approximation $R_{\text{pert}}^{(\ell)}(s)$ (C.3) contains infrared unphysical singularities, becomes inapplicable for the energies $\sqrt{s} < \Lambda \exp(\pi/2) \simeq 4.81\Lambda$, and slowly converges even at relatively high energies, see discussion of these issues in Chapter 6 for the details.

The explicit expressions for the coefficients δ_j up to the twelfth order ($1 \leq j \leq 12$) are presented in the following. In these equations $B_j = \beta_j/\beta_0^{j+1}$ denotes the ratio of the β function perturbative expansion coefficients (see Section 2.1). The numerical values of the coefficients δ_j and r_j are listed in Tables 6.1 and 6.2, respectively.

First-order coefficient:

$$\delta_1 = 0, \tag{C.4}$$

Second-order coefficient:

$$\delta_2 = 0, \tag{C.5}$$

Third-order coefficient:

$$\delta_3 = \frac{\pi^2}{3} d_1, \tag{C.6}$$

Fourth-order coefficient:

$$\delta_4 = \frac{\pi^2}{3} \left(\frac{5}{2} d_1 B_1 + 3 d_2 \right), \tag{C.7}$$

Fifth-order coefficient:

$$\delta_5 = \frac{\pi^2}{3} \left[\frac{3}{2} d_1 \left(B_1^2 + 2 B_2 \right) + 7 d_2 B_1 + 6 d_3 \right] - \frac{\pi^4}{5} d_1, \tag{C.8}$$

Sixth-order coefficient:

$$\delta_6 = \frac{\pi^2}{3} \left[\frac{7}{2}d_1 \left(B_1B_2 + B_3\right) + 4d_2 \left(B_1^2 + 2B_2\right) + \frac{27}{2}d_3B_1 + 10d_4 \right]$$
$$- \frac{\pi^4}{5} \left(\frac{77}{12}d_1B_1 + 5d_2 \right), \tag{C.9}$$

Seventh-order coefficient:

$$\delta_7 = \frac{\pi^2}{3} \left[4d_1 \left(B_1B_3 + \frac{1}{2}B_2^2 + B_4\right) + 9d_2 \left(B_1B_2 + B_3\right) \right.$$
$$\left. + \frac{15}{2}d_3 \left(B_1^2 + 2B_2\right) + 22d_4B_1 + 15d_5 \right]$$
$$- \frac{\pi^4}{5} \left[\frac{5}{6}d_1 \left(17B_1^2 + 12B_2\right) + \frac{57}{2}d_2B_1 + 15d_3 \right] + \frac{\pi^6}{7}d_1, \tag{C.10}$$

Eighth-order coefficient:

$$\delta_8 = \frac{\pi^2}{3} \left[\frac{9}{2}d_1 \left(B_1B_4 + B_2B_3 + B_5\right) + 10d_2 \left(B_1B_3 + \frac{1}{2}B_2^2 + B_4\right) \right.$$
$$\left. + \frac{33}{2}d_3 \left(B_1B_2 + B_3\right) + 12d_4 \left(B_1^2 + 2B_2\right) + \frac{65}{2}d_5B_1 + 21d_6 \right]$$
$$- \frac{\pi^4}{5} \left[\frac{15}{8}d_1 \left(7B_1^3 + 22B_1B_2 + 8B_3\right) + \frac{5}{12}d_2 \left(139B_1^2 + 96B_2\right) \right.$$
$$\left. + \frac{319}{4}d_3B_1 + 35d_4 \right] + \frac{\pi^6}{7} \left(\frac{223}{20}d_1B_1 + 7d_2 \right), \tag{C.11}$$

Ninth-order coefficient:

$$\delta_9 = \frac{\pi^2}{3} \left[5d_1 \left(B_1B_5 + B_2B_4 + \frac{1}{2}B_3^2 + B_6\right) + 11d_2 \left(B_1B_4 + B_2B_3 + B_5\right) \right.$$
$$+ 18d_3 \left(B_1B_3 + \frac{1}{2}B_2^2 + B_4\right) + 26d_4 \left(B_1B_2 + B_3\right)$$
$$\left. + \frac{35}{2}d_5 \left(B_1^2 + 2B_2\right) + 45d_6B_1 + 28d_7 \right]$$
$$- \frac{\pi^4}{5} \left[\frac{5}{12}d_1 \left(\frac{21}{2}B_1^4 + 131B_1^2B_2 + 141B_1B_3 + 68B_2^2 + 52B_4 \right) \right.$$
$$+ \frac{55}{12}d_2 \left(11B_1^3 + 34B_1B_2 + 12B_3\right) + \frac{15}{4}d_3 \left(41B_1^2 + 28B_2\right)$$
$$\left. + \frac{533}{3}d_4B_1 + 70d_5 \right] + \frac{\pi^6}{7} \left[\frac{7}{36}d_1 \left(241B_1^2 + 108B_2\right) \right.$$
$$\left. + \frac{341}{5}d_2B_1 + 28d_3 \right] - \frac{\pi^8}{9}d_1, \tag{C.12}$$

Tenth-order coefficient:

$$\delta_{10} = \frac{\pi^2}{3}\left[\frac{11}{2}d_1\left(B_1B_6 + B_2B_5 + B_3B_4 + B_7\right)\right.$$

$$+ 12d_2\left(B_1B_5 + B_2B_4 + \frac{1}{2}B_3^2 + B_6\right) + \frac{39}{2}d_3\left(B_1B_4 + B_2B_3 + B_5\right)$$

$$+ 28d_4\left(B_1B_3 + \frac{1}{2}B_2^2 + B_4\right) + \frac{75}{2}d_5\left(B_1B_2 + B_3\right)$$

$$\left. + 24d_6\left(B_1^2 + 2B_2\right) + \frac{119}{2}d_7B_1 + 36d_8\right]$$

$$- \frac{\pi^4}{5}\left[\frac{11}{24}d_1\left(51B_1^3B_2 + 163B_1^2B_3 + 158B_1B_2^2 + 178B_1B_4\right.\right.$$

$$\left.+ 168B_2B_3 + 66B_5\right) + d_2\left(16B_1^4 + 197B_1^2B_2 + 207B_1B_3\right.$$

$$\left.+ 101B_2^2 + 74B_4\right) + \frac{13}{8}d_3\left(79B_1^3 + 242B_1B_2 + 84B_3\right)$$

$$\left. + \frac{7}{6}d_4\left(283B_1^2 + 192B_2\right) + \frac{1375}{4}d_5B_1 + 126d_6\right]$$

$$+ \frac{\pi^6}{7}\left[\frac{77}{180}d_1\left(229B_1^3 + 378B_1B_2 + 90B_3\right)\right.$$

$$\left. + \frac{7}{20}d_2\left(743B_1^2 + 320B_2\right) + \frac{2509}{10}d_3B_1 + 84d_4\right]$$

$$- \frac{\pi^8}{9}\left[\frac{4609}{280}d_1B_1 + 9d_2\right], \tag{C.13}$$

Eleventh-order coefficient:

$$\delta_{11} = \frac{\pi^2}{3}\left[3d_1\left(2B_1B_7 + 2B_2B_6 + 2B_3B_5 + B_4^2 + 2B_8\right)\right.$$

$$+ 13d_2\left(B_1B_6 + B_2B_5 + B_3B_4 + B_7\right) + \frac{21}{2}d_3\left(2B_1B_5 + 2B_2B_4 + B_3^2 + 2B_6\right)$$

$$+ 30d_4\left(B_1B_4 + B_2B_3 + B_5\right) + 20d_5\left(2B_1B_3 + B_2^2 + 2B_4\right)$$

$$\left. + 51d_6\left(B_1B_2 + B_3\right) + \frac{63}{2}d_7\left(B_1^2 + 2B_2\right) + 76d_8B_1 + 45d_9\right]$$

$$- \frac{\pi^4}{5}\left[\frac{1}{4}d_1\left(181B_1^2B_2^2 + 764B_1B_2B_3 + 124B_1^3B_3 + 402B_1^2B_4\right.\right.$$

$$\left.+ 442B_1B_5 + 124B_2^3 + 412B_2B_4 + 201B_3^2 + 164B_6\right)$$

$$+ \frac{65}{12}d_2\left(15B_1^3B_2 + 46B_1B_2^2 + 47B_1^2B_3 + 50B_1B_4 + 48B_2B_3 + 18B_5\right)$$

$$+ \frac{35}{8}d_3\left(9B_1^4 + 110B_1^2B_2 + 114B_1B_3 + 56B_2^2 + 40B_4\right) + \frac{5}{2}d_4\left(107B_1^3\right.$$

$$
+326B_1B_2 + 112B_3) + \frac{5}{3}d_5 \left(373B_1^2 + 252B_2\right) + \frac{1207}{2}d_6B_1 + 210d_7 \Big]
$$

$$
+ \frac{\pi^6}{7} \left[\frac{7}{120}d_1 \left(1875B_1^4 + 8144B_1^2B_2 + 4744B_1B_3 + 2200B_2^2 + 1160B_4\right) \right.
$$

$$
+ \frac{91}{90}d_2 \left(499B_1^3 + 798B_1B_2 + 180B_3\right) + \frac{7}{60}d_3 \left(7687B_1^2 + 3240B_2\right)
$$

$$
+ \frac{2131}{3}d_4B_1 + 210d_5 \Big]
$$

$$
- \frac{\pi^8}{9} \left[\frac{1}{560}d_1 \left(59753B_1^2 + 20160B_2\right) + \frac{3601}{28}d_2B_1 + 45d_3 \right] + \frac{\pi^{10}}{11}d_1,
$$

$$
\text{(C.14)}
$$

Twelfth-order coefficient:

$$
\delta_{12} = \frac{\pi^2}{3} \left[\frac{13}{2}d_1 \left(B_1B_8 + B_2B_7 + B_3B_6 + B_4B_5 + B_9\right) \right.
$$

$$
+ 7d_2 \left(2B_1B_7 + 2B_2B_6 + 2B_3B_5 + B_4^2 + 2B_8\right)
$$

$$
+ \frac{45}{2}d_3 \left(B_1B_6 + B_2B_5 + B_3B_4 + B_7\right) + 16d_4 \left(2B_1B_5 + 2B_2B_4 + B_3^2 + 2B_6\right)
$$

$$
+ \frac{85}{2}d_5 \left(B_1B_4 + B_2B_3 + B_5\right) + 27d_6 \left(2B_1B_3 + B_2^2 + 2B_4\right) + \frac{133}{2}d_7 \left(B_3 + B_1B_2\right)
$$

$$
+ 40d_8 \left(B_1^2 + 2B_2\right) + \frac{189}{2}d_9B_1 + 55d_{10} \Big] - \frac{\pi^4}{5} \left[\frac{65}{24}d_1 \left(14B_1B_2^3 + 43B_1^3B_2B_3 \right. \right.
$$

$$
+ 45B_1B_3^2 + 92B_1B_2B_4 + 15B_1^3B_4 + 49B_1^2B_5 + 54B_1B_6 + 44B_2^2B_3 + 50B_2B_5
$$

$$
+ 48B_3B_4 + 20B_7\big) + \frac{7}{12}d_2 \left(259B_1^2B_2^2 + 176B_1^3B_3 + 1076B_1B_2B_3 + 558B_1^2B_4 \right.
$$

$$
+ 598B_1B_5 + 176B_2^3 + 568B_2B_4 + 279B_3^2 + 216B_6\big) + \frac{15}{8}d_3 \left(103B_1^3B_2 + 314B_1B_2^2 \right.
$$

$$
+ 319B_1^2B_3 + 334B_1B_4 + 324B_2B_3 + 118B_5\big) + \frac{40}{3}d_4 \left(6B_1^4 + 73B_1^2B_2 + 75B_1B_3 \right.
$$

$$
+ 37B_2^2 + 26B_4\big) + \frac{85}{24}d_5 \left(139B_1^3 + 422B_1B_2 + 144B_3\right) + \frac{45}{4}d_6 \left(95B_1^2 + 64B_2\right)
$$

$$
+ \frac{11837}{12}d_7B_1 + 330d_8 \Big] + \frac{\pi^6}{7} \left[\frac{91}{720}d_1 \left(495B_1^5 + 5320B_1^3B_2 + 5696B_1B_2^2 \right. \right.
$$

$$
+ 6092B_1^2B_3 + 3652B_1B_4 + 3240B_2B_3 + 900B_5\big) + \frac{7}{360}d_2 \left(27295B_1^4 \right.
$$

$$
+ 115508B_1^2B_2 + 64278B_1B_3 + 30480B_2^2 + 14880B_4\big) + \frac{7}{4}d_3 \left(944B_1^3 + 1483B_1B_2 \right.
$$

$$
+ 324B_3\big) + \frac{14}{45}d_4 \left(7789B_1^2 + 3240B_2\right) + \frac{20417}{12}d_5B_1 + 462d_6 \Big]
$$

$$
- \frac{\pi^8}{9} \left[\frac{13}{2016}d_1 \left(56837B_1^3 + 65556B_1B_2 + 12096B_3\right) + \frac{1}{80}d_2 \left(60079B_1^2 \right. \right.
$$

$$
+ 19200B_2\big) + \frac{32891}{56}d_3B_1 + 165d_4 \Big] + \frac{\pi^{10}}{11} \left(\frac{55991}{2520}d_1B_1 + 11d_2 \right).
$$

$$
\text{(C.15)}
$$

Bibliography

[1] M. Gell-Mann, A schematic model of baryons and mesons, Phys. Lett. 8 (1964) 214.

[2] G. Zweig, An SU(3) model for strong interaction symmetry and its breaking, Report CERN-TH-401, 1964.

[3] G. Zweig, An SU(3) model for strong interaction symmetry and its breaking (version II), Report CERN-TH-412, 1964.

[4] A. Petermann, Proprietes de l'etrangete et une formule de masse pour les mesons vectoriels, Nucl. Phys. 63 (1965) 349.

[5] A. de Rujula, Who invented quarks?, CERN Courier 54 (4) (2014) 35.

[6] O.W. Greenberg, Spin and unitary spin independence in a paraquark model of baryons and mesons, Phys. Rev. Lett. 13 (1964) 598.

[7] M.Y. Han, Y. Nambu, Three triplet model with double SU(3) symmetry, Phys. Rev. 139 (1965) B1006.

[8] Y. Nambu, A systematics of hadrons in subnuclear physics, in: A. de Shalit, H. Feshbach, L. van Hove (Eds.), Preludes in Theoretical Physics, North Holland, Amsterdam, 1966, p. 133.

[9] B.V. Struminsky, Magnetic moments of baryons in the quark model, Report JINR P-1939, Dubna, 1965.

[10] N.N. Bogoliubov, B.V. Struminsky, A.N. Tavkhelidze, On composite models in the theory of elementary particles, Report JINR D-1968, Dubna, 1965.

[11] A. Pais, in: A. Zichichi (Ed.), Recent Developments in Particle Symmetries, Academic Press, New York, 1966, p. 406.

[12] O.W. Greenberg, From Wigner's supermultiplet theory to Quantum Chromodynamics, Acta Phys. Hung. A 19 (2004) 353, arXiv:hep-ph/0212174.

[13] O.W. Greenberg, Discovery of the color degree of freedom in particle physics: a personal perspective, arXiv:0803.0992 [physics.hist-ph].

[14] O.W. Greenberg, The color charge degree of freedom in particle physics, in: D. Greenberger, K. Hentschel, F. Weinert (Eds.), Compendium of Quantum Physics, Springer-Verlag, Berlin/Heidelberg, 2009, p. 109, arXiv:0805.0289 [physics.hist-ph].

[15] G. 't Hooft, Report at the Colloquium on renormalization of Yang-Mills fields and applications to particle physics, 19–23 June, Marseille, France, 1972.

[16] D.J. Gross, F. Wilczek, Ultraviolet behavior of nonabelian gauge theories, Phys. Rev. Lett. 30 (1973) 1343.

[17] H.D. Politzer, Reliable perturbative results for strong interactions? Phys. Rev. Lett. 30 (1973) 1346.

[18] D.J. Gross, F. Wilczek, Asymptotically free gauge theories 1, Phys. Rev. D 8 (1973) 3633.

[19] D.J. Gross, F. Wilczek, Asymptotically free gauge theories 2, Phys. Rev. D 9 (1974) 980.

[20] G. 't Hooft, The birth of asymptotic freedom, Nucl. Phys. B 254 (1985) 11.

[21] G. 't Hooft, When was asymptotic freedom discovered? Or the rehabilitation of Quantum Field Theory, Nucl. Phys. B (Proc. Suppl.) 74 (1999) 413, arXiv:hep-th/9808154.

[22] T. Goto, Relativistic quantum mechanics of one-dimensional mechanical continuum and subsidiary condition of dual resonance model, Prog. Theor. Phys. 46 (1971) 1560.

[23] Y. Nambu, Strings, monopoles and gauge fields, Phys. Rev. D 10 (1974) 4262.

[24] R. Andreo, F. Rohrlich, A string model of mesons, Nucl. Phys. B 115 (1976) 521.

[25] K. Kamimura, Massive string model of mesons and baryons, Prog. Theor. Phys. 62 (1979) 508.

[26] B.M. Barbashov, V.V. Nesterenko, Introduction to the Relativistic String Theory, World Scientific, Singapore, 1990, 264 pp.

[27] J.L. Richardson, The heavy quark potential and the Υ, J/ψ systems, Phys. Lett. B 82 (1979) 272.

[28] A.A. Bykov, I.M. Dremin, A.V. Leonidov, Potential models of quarkonium, Sov. Phys. Usp. 27 (1984) 321.

[29] A.W. Hendry, D.B. Lichtenberg, Properties of hadrons in the quark model, Fortsch. Phys. 33 (1985) 139.

[30] W. Lucha, F.F. Schoberl, D. Gromes, Bound states of quarks, Phys. Rep. 200 (1991) 127.

[31] W. Lucha, F.F. Schoberl, Effective potential models for hadrons, arXiv:hep-ph/9601263.

[32] N. Brambilla, A. Vairo, Quark confinement and the hadron spectrum, arXiv:hep-ph/9904330.

[33] P. Hasenfratz, J. Kuti, The quark bag model, Phys. Rep. 40 (1978) 75.

[34] C.E. DeTar, J.F. Donoghue, Bag models of hadrons, Ann. Rev. Nucl. Part. Sci. 33 (1983) 235.

[35] S.L. Adler, T. Piran, Relaxation methods for gauge field equilibrium equations, Rev. Mod. Phys. 56 (1984) 1.

[36] H.M. Fried, Modern Functional Quantum Field Theory: Summing Feynman Graphs, World Scientific, Singapore, 2014, 266 pp.

[37] H.M. Fried, Y. Gabellini, T. Grandou, Y.-M. Sheu, Gauge invariant summation of all QCD virtual gluon exchanges, Eur. Phys. J. C 65 (2010) 395, arXiv:0903.2644 [hep-th].

[38] H.M. Fried, T. Grandou, Y.-M. Sheu, A new approach to analytic, non-perturbative and gauge-invariant QCD, Ann. Phys. 327 (2012) 2666, arXiv:1204.2038 [hep-ph].

[39] H.M. Fried, Y. Gabellini, T. Grandou, Y.-M. Sheu, Analytic, non-perturbative, gauge-invariant Quantum Chromodynamics: nucleon scattering and binding potentials, Ann. Phys. 338 (2013) 107, arXiv:1203.6137 [hep-ph].

[40] H.M. Fried, T. Grandou, Y.-M. Sheu, Non-perturbative QCD amplitudes in quenched and eikonal approximations, Ann. Phys. 344 (2014) 78, arXiv:1207.5017 [hep-th].

[41] H.M. Fried, P.H. Tsang, Y. Gabellini, T. Grandou, Y.-M. Sheu, An exact, finite, gauge-invariant, non-perturbative approach to QCD renormalization, Ann. Phys. 359 (2015) 1, arXiv:1412.2072 [hep-th].

[42] U. Gursoy, E. Kiritsis, Exploring improved holographic theories for QCD: part I, JHEP 802 (2008) 32, arXiv:0707.1324 [hep-th].

[43] U. Gursoy, E. Kiritsis, F. Nitti, Exploring improved holographic theories for QCD: part II, JHEP 802 (2008) 19, arXiv:0707.1349 [hep-th].

[44] Y. Kim, I.J. Shin, T. Tsukioka, Holographic QCD: past, present, and future, Prog. Part. Nucl. Phys. 68 (2013) 55, arXiv:1205.4852 [hep-ph].

[45] S.J. Brodsky, G.F. de Teramond, H.G. Dosch, J. Erlich, Light-front holographic QCD and emerging confinement, Phys. Rep. 584 (2015) 1, arXiv:1407.8131 [hep-ph].

[46] R. Gupta, Introduction to lattice QCD, arXiv:hep-lat/9807028.

[47] G.S. Bali, QCD forces and heavy quark bound states, Phys. Rep. 343 (2001) 1, arXiv:hep-ph/0001312.

[48] A.M. Green (Ed.), Hadronic Physics from Lattice QCD, World Scientific, Singapore, 2004, 367 pp.

[49] T. DeGrand, C.E. Detar, Lattice Methods for Quantum Chromodynamics, World Scientific, Singapore, 2006, 364 pp.

[50] C. Gattringer, C.B. Lang, Quantum Chromodynamics on the lattice, Lect. Notes Phys. 788 (2010) 1.

[51] H.J. Rothe, Lattice gauge theories: an introduction, World Sci. Lect. Notes Phys. 82 (2012) 1.

[52] P. Hagler, Hadron structure from lattice Quantum Chromodynamics, Phys. Rep. 490 (2010) 49, arXiv:0912.5483 [hep-lat].

[53] G. Colangelo, et al. [FLAG Working Group], Review of lattice results concerning low energy particle physics, Eur. Phys. J. C 71 (2011) 1695, arXiv:1011.4408 [hep-lat].

[54] S. Aoki, et al. [FLAG Working Group], Review of lattice results concerning low-energy particle physics, Eur. Phys. J. C 74 (2014) 2890, arXiv:1310.8555 [hep-lat].

[55] N.N. Bogoliubov, B.V. Medvedev, M.K. Polivanov, Problems in the Theory of Dispersion Relations, Institute for Advanced Study, Princeton, NJ, 1958, 200 pp.

[56] R. Hagedorn, Introduction to Field Theory and Dispersion Relations, Pergamon Press, Oxford, 1964, 127 pp.

[57] G. Barton, Introduction to Dispersion Techniques in Field Theory, Benjamin, New York, 1965, 242 pp.

[58] D.V. Shirkov, V.V. Serebryakov, V.A. Mescheryakov, Dispersion Theories of Strong Interactions at Low Energy, North-Holland, Amsterdam, 1969, 362 pp.

[59] N.M. Queen, G. Violini, Dispersion Theory in High Energy Physics, Macmillan, London, 1974, 202 pp.

[60] A.A. Logunov, L.D. Soloviev, A.N. Tavkhelidze, Dispersion sum rules and high-energy scattering, Phys. Lett. B 24 (1967) 181.

[61] J.J. Sakurai, Duality in $e^+e^- \to$ hadrons?, Phys. Lett. B 46 (1973) 207.

[62] K.G. Chetyrkin, N.V. Krasnikov, A.N. Tavkhelidze, Finite energy sum rules for the cross–section of e^+e^- annihilation into hadrons in QCD, Phys. Lett. B 76 (1978) 83.

[63] N.V. Krasnikov, A.A. Pivovarov, N.N. Tavkhelidze, The use of finite energy sum rules for the description of the hadronic properties of QCD, Z. Phys. C 19 (1983) 301.

[64] H. Leutwyler, On the foundations of chiral perturbation theory, Ann. Phys. 235 (1994) 165, arXiv:hep-ph/9311274.

[65] A. Pich, Chiral perturbation theory, Rept. Prog. Phys. 58 (1995) 563, arXiv:hep-ph/9502366.

[66] G. Ecker, Chiral perturbation theory, Prog. Part. Nucl. Phys. 35 (1995) 1, arXiv:hep-ph/9501357.

[67] S. Scherer, M.R. Schindler, A primer for chiral perturbation theory, Lect. Notes Phys. 830 (2012) 1.

[68] F. Guerrero, A. Pich, Effective field theory description of the pion form-factor, Phys. Lett. B 412 (1997) 382, arXiv:hep-ph/9707347.

[69] A. Pich, J. Portoles, The vector form-factor of the pion from unitarity and analyticity: a model independent approach, Phys. Rev. D 63 (2001) 93005, arXiv:hep-ph/0101194.

[70] D.G. Dumm, P. Roig, Dispersive representation of the pion vector form factor in $\tau \to \pi\pi\nu_\tau$ decays, Eur. Phys. J. C 73 (2013) 2528, arXiv:1301.6973 [hep-ph].

[71] P. Roig, A. Guevara, G.L. Castro, $VV'P$ form factors in resonance chiral theory and the $\pi - \eta - \eta'$ light-by-light contribution to the muon $g - 2$, Phys. Rev. D 89 (2014) 73016, arXiv:1401.4099 [hep-ph].

[72] V. Bernard, E. Passemar, Matching chiral perturbation theory and the dispersive representation of the scalar $K\pi$ form-factor, Phys. Lett. B 661 (2008) 95, arXiv:0711.3450 [hep-ph].

[73] V. Bernard, M. Oertel, E. Passemar, J. Stern, Dispersive representation and shape of the K_3 form factors: robustness, Phys. Rev. D 80 (2009) 34034, arXiv:0903.1654 [hep-ph].

[74] G. Colangelo, E. Passemar, P. Stoffer, A dispersive treatment of K_4 decays, Eur. Phys. J. C 75 (2015) 172, arXiv:1501.05627 [hep-ph].

[75] R. Garcia-Martin, R. Kaminski, J.R. Pelaez, J.R. de Elvira, Precise determination of the $f_0(600)$ and $f_0(980)$ pole parameters from a dispersive data analysis, Phys. Rev. Lett. 107 (2011) 72001, arXiv:1107.1635 [hep-ph].

[76] R. Garcia-Martin, R. Kaminski, J.R. Pelaez, J.R. de Elvira, F.J. Yndurain, The pion-pion scattering amplitude: improved analysis with once subtracted Roy-like equations up to 1100 MeV, Phys. Rev. D 83 (2011) 74004, arXiv:1102.2183 [hep-ph].

[77] G. Colangelo, M. Hoferichter, A. Nyffeler, M. Passera, P. Stoffer, Remarks on higher-order hadronic corrections to the muon $g - 2$, Phys. Lett. B 735 (2014) 90, arXiv:1403.7512 [hep-ph].

[78] G. Colangelo, M. Hoferichter, B. Kubis, M. Procura, P. Stoffer, Towards a data-driven analysis of hadronic light-by-light scattering, Phys. Lett. B 738 (2014) 6, arXiv:1408.2517 [hep-ph].

[79] G. Colangelo, M. Hoferichter, M. Procura, P. Stoffer, Dispersive approach to hadronic light-by-light scattering, JHEP 1409 (2014) 91, arXiv:1402.7081 [hep-ph].

[80] G. Colangelo, M. Hoferichter, M. Procura, P. Stoffer, Dispersion relation for hadronic light-by-light scattering: theoretical foundations, JHEP 1509 (2015) 74, arXiv:1506.01386 [hep-ph].

[81] H.J. Bhabha, The scattering of positrons by electrons with exchange on Dirac's theory of the positron, Proc. R. Soc. Lond. A 154 (1936) 195.

[82] R.P. Feynman, The theory of positrons, Phys. Rev. 76 (1949) 749.

[83] R.P. Feynman, Space-time approach to Quantum Electrodynamics, Phys. Rev. 76 (1949) 769.

[84] K.A. Olive, et al. [Particle Data Group Collaboration], Review of particle physics, Chin. Phys. C 38 (2014) 90001.

[85] H. Burkhardt, New numerical analysis of the hadronic vacuum polarization, TASSO-NOTE-192; relevant computer program is available at: http://hbu.web.cern.ch/hbu.

[86] H. Burkhardt, B. Pietrzyk, Update of the hadronic contribution to the QED vacuum polarization, Phys. Lett. B 356 (1995) 398.

[87] H. Burkhardt, B. Pietrzyk, Update of the hadronic contribution to the QED vacuum polarization, Phys. Lett. B 513 (2001) 46.

[88] H. Burkhardt, B. Pietrzyk, Low energy hadronic contribution to the QED vacuum polarization, Phys. Rev. D 72 (2005) 57501, arXiv:hep-ph/0506323.

[89] H. Burkhardt, B. Pietrzyk, Recent BES measurements and the hadronic contribution to the QED vacuum polarization, Phys. Rev. D 84 (2011) 37502, arXiv:1106.2991 [hep-ex].

[90] F. Jegerlehner, Hadronic contributions to electroweak parameter shifts: a detailed analysis, Z. Phys. C 32 (1986) 195.

[91] S. Eidelman, F. Jegerlehner, Hadronic contributions to $g - 2$ of the leptons and to the effective fine structure constant $\alpha(M_Z^2)$, Z. Phys. C 67 (1995) 585, arXiv:hep-ph/9502298.

[92] F. Jegerlehner, The running fine structure constant $\alpha(E)$ via the Adler function, Nucl. Phys. B (Proc. Suppl.) 181 (2008) 135, arXiv:0807.4206 [hep-ph].

[93] F. Jegerlehner, Electroweak effective couplings for future precision experiments, Nuovo Cim. C 34S1 (2011) 31, arXiv:1107.4683 [hep-ph]; relevant computer package "alphaQED" is available at: http://www-com.physik.hu-berlin.de/~fjeger/.

[94] K. Hagiwara, A.D. Martin, D. Nomura, T. Teubner, Predictions for $g - 2$ of the muon and $\alpha_{\text{QED}}(M_Z^2)$, Phys. Rev. D 69 (2004) 93003, arXiv:hep-ph/0312250.

[95] K. Hagiwara, A.D. Martin, D. Nomura, T. Teubner, Improved predictions for $g - 2$ of the muon and $\alpha_{\text{QED}}(M_Z^2)$, Phys. Lett. B 649 (2007) 173, arXiv:hep-ph/0611102.

[96] K. Hagiwara, R. Liao, A.D. Martin, D. Nomura, T. Teubner, $(g - 2)_\mu$ and $\alpha(M_Z^2)$ re-evaluated using new precise data, J. Phys. G 38 (2011) 85003, arXiv:1105.3149 [hep-ph].

[97] T. Teubner, K. Hagiwara, R. Liao, A.D. Martin, D. Nomura, $g - 2$ and $\alpha(M_Z^2)$: status of the standard model predictions, Nucl. Phys. B (Proc. Suppl.) 225 (2012) 282.

[98] S. Actis, et al. [Working Group on Radiative Corrections and Monte Carlo Generators for Low Energies Collaboration], M.C. generators for low energies Collaboration, Quest for precision in hadronic cross sections at low energy: Monte Carlo tools vs. experimental data, Eur. Phys. J. C 66 (2010) 585, arXiv:0912.0749 [hep-ph].

[99] R.P. Feynman, Photon-Hadron Interactions, Benjamin, Reading, MA, 1972, 282 pp.

[100] S.L. Adler, Some simple vacuum-polarization phenomenology: $e^+e^- \rightarrow$ hadrons; the muonic-atom X-ray discrepancy and $(g_\mu - 2)$, Phys. Rev. D 10 (1974) 3714.

[101] F.J. Gilman, Theory of electron-positron annihilation into hadrons, Report SLAC-PUB-1650, 1975.

[102] E.C. Poggio, H.R. Quinn, S. Weinberg, Smearing the quark model, Phys. Rev. D 13 (1976) 1958.

[103] N. Cabibbo, G. Parisi, M. Testa, Hadron production in e^+e^- collisions, Lett. Nuovo Cim. 4 (1970) 35.

[104] A. De Rujula, H. Georgi, Counting quarks in e^+e^- annihilation, Phys. Rev. D 13 (1976) 1296.

[105] A.V. Radyushkin, Optimized Λ-parametrization for the QCD running coupling constant in spacelike and timelike regions, JINR Rapid Commun. 78 (1996) 96, Report JINR 2-82-159 (1982).

[106] N.V. Krasnikov, A.A. Pivovarov, The influence of the analytic continuation effects on the value of the QCD scale parameter Λ extracted from charmonium and upsilonium hadron decays, Phys. Lett. B 116 (1982) 168.

[107] J.D. Bjorken, Two topics in Quantum Chromodynamics, Report SLAC-PUB-5103, 1989.

[108] A.L. Kataev, V.V. Starshenko, Estimates of the higher order QCD corrections to $R(s)$, R_τ, and deep inelastic scattering sum rules, Mod. Phys. Lett. A 10 (1995) 235, arXiv:hep-ph/9502348.

[109] L.V. Ahlfors, Complex Analysis: An Introduction to the Theory of Analytic Functions of One Complex Variable, third ed., McGraw-Hill, New York, 1979, 331 pp.

[110] E. Freitag, R. Busam, Complex Analysis, second ed., Springer, Berlin, 2009, 532 pp.

[111] J. Bruna, J. Cufi, Complex Analysis, European Mathematical Society, Zurich, 2013, 564 pp.

[112] B.E. Lautrup, A. Peterman, E. de Rafael, Recent developments in the comparison between theory and experiments in Quantum Electrodynamics, Phys. Rep. 3 (1972) 193.

[113] J. Calmet, S. Narison, M. Perrottet, E. de Rafael, The anomalous magnetic moment of the muon: a review of the theoretical contributions, Rev. Mod. Phys. 49 (1977) 21.

[114] M. Knecht, The anomalous magnetic moment of the muon: a theoretical introduction, Lect. Notes Phys. 629 (2004) 37, arXiv:hep-ph/0307239.

[115] M. Davier, W.J. Marciano, The theoretical prediction for the muon anomalous magnetic moment, Ann. Rev. Nucl. Part. Sci. 54 (2004) 115.

[116] M. Passera, The standard model prediction of the muon anomalous magnetic moment, J. Phys. G 31 (2005) R75, hep-ph/0411168.

[117] F. Jegerlehner, A. Nyffeler, The muon $g - 2$, Phys. Rep. 477 (2009) 1, arXiv:0902.3360 [hep-ph].

[118] J.P. Miller, E. de Rafael, B.L. Roberts, D. Stockinger, Muon $(g - 2)$: experiment and theory, Ann. Rev. Nucl. Part. Sci. 62 (2012) 237.

[119] F. Jegerlehner, R. Szafron, $\rho^0 - \gamma$ mixing in the neutral channel pion form factor F_π^e and its role in comparing $e^+ e^-$ with τ spectral functions, Eur. Phys. J. C 71 (2011) 1632, arXiv:1101.2872 [hep-ph].

[120] M. Davier, A. Hocker, B. Malaescu, Z. Zhang, Reevaluation of the hadronic contributions to the muon $g - 2$ and to $\alpha(M_Z^2)$, Eur. Phys. J. C 71 (2011) 1515, 72 (2012) 1874(E); arXiv:1010.4180 [hep-ph].

[121] F. Jegerlehner, Leading-order hadronic contribution to the electron and muon $g - 2$, EPJ Web Conf. 118 (2016) 1016, arXiv:1511.04473 [hep-ph].

[122] R. Barate, et al. [ALEPH Collaboration], Measurement of the spectral functions of vector current hadronic τ decays, Z. Phys. C 76 (1997) 15.

[123] R. Barate, et al. [ALEPH Collaboration], Measurement of the axial-vector τ spectral functions and determination of $\alpha_s(M_\tau^2)$ from hadronic τ decays, Eur. Phys. J. C 4 (1998) 409.

[124] S. Schael, et al. [ALEPH Collaboration], Branching ratios and spectral functions of τ decays: final ALEPH measurements and physics implications, Phys. Rep. 421 (2005) 191, arXiv:hep-ex/0506072.

[125] K. Ackerstaff, et al. [OPAL Collaboration], Measurement of the strong coupling constant α_s and the vector and axial-vector spectral functions in hadronic τ decays, Eur. Phys. J. C 7 (1999) 571, arXiv:hep-ex/9808019.

[126] M. Davier, A. Hocker, Z. Zhang, The physics of hadronic τ decays, Rev. Mod. Phys. 78 (2006) 1043, arXiv:hep-ph/0507078.

[127] M. Davier, S. Descotes-Genon, A. Hocker, B. Malaescu, Z. Zhang, The determination of α_s from τ decays revisited, Eur. Phys. J. C 56 (2008) 305, arXiv:0803.0979 [hep-ph].

[128] M. Davier, A. Hocker, B. Malaescu, C.Z. Yuan, Z. Zhang, Update of the ALEPH non-strange spectral functions from hadronic τ decays, Eur. Phys. J. C 74 (2014) 2803, arXiv:1312.1501 [hep-ex].

[129] D. Boito, O. Cata, M. Golterman, M. Jamin, K. Maltman, J. Osborne, S. Peris, A new determination of α_s from hadronic τ decays, Phys. Rev. D 84 (2011) 113006, arXiv:1110.1127 [hep-ph].

[130] D. Boito, M. Golterman, M. Jamin, A. Mahdavi, K. Maltman, J. Osborne, S. Peris, An updated determination of α_s from τ decays, Phys. Rev. D 85 (2012) 93015, arXiv:1203.3146 [hep-ph].

[131] D. Boito, M. Golterman, K. Maltman, J. Osborne, S. Peris, Strong coupling from the revised ALEPH data for hadronic τ decays, Phys. Rev. D 91 (2015) 34003, arXiv:1410.3528 [hep-ph].

[132] M. Davier, S. Eidelman, A. Hocker, Z. Zhang, Confronting spectral functions from $e^+ e^-$ annihilation and τ decays: consequences for the muon magnetic moment, Eur. Phys. J. C 27 (2003) 497, arXiv:hep-ph/0208177.

[133] M. Davier, S. Eidelman, A. Hocker, Z. Zhang, Updated estimate of the muon magnetic moment using revised results from $e^+ e^-$ annihilation, Eur. Phys. J. C 31 (2003) 503, arXiv:hep-ph/0308213.

[134] M. Davier, A. Hocker, G.L. Castro, B. Malaescu, X.H. Mo, G.T. Sanchez, P. Wang, C.Z. Yuan, Z. Zhang, The discrepancy between τ and $e^+ e^-$ spectral functions revisited and the consequences for the muon magnetic anomaly, Eur. Phys. J. C 66 (2010) 127, arXiv:0906.5443 [hep-ph].

[135] R.G. Moorhouse, M.R. Pennington, G.G. Ross, What can asymptotic freedom say about $e^+e^- \to$ hadrons?, Nucl. Phys. B 124 (1977) 285.

[136] M.R. Pennington, G.G. Ross, Perturbative QCD for timelike processes: what is the best expansion parameter?, Phys. Lett. B 102 (1981) 167.

[137] M.R. Pennington, R.G. Roberts, G.G. Ross, How to continue the predictions of perturbative QCD from the spacelike region where they are derived to the timelike regime where experiments are performed, Nucl. Phys. B 242 (1984) 69.

[138] A.I. Sanda, A nonperturbative determination of $\alpha(q^2)$ and its experimental implications, Phys. Rev. Lett. 42 (1979) 1658.

[139] A.A. Pivovarov, Renormalization group summation of perturbative series in timelike momentum region, Nuovo Cim. A 105 (1992) 813.

[140] S. Eidelman, F. Jegerlehner, A.L. Kataev, O. Veretin, Testing nonperturbative strong interaction effects via the Adler function, Phys. Lett. B 454 (1999) 369, arXiv:hep-ph/9812521.

[141] A.L. Kataev, Adler function from $R_{e^+e^-}(s)$ measurements: experiments versus QCD theory, arXiv:hep-ph/9906534.

[142] G. Leibbrandt, Introduction to the technique of dimensional regularization, Rev. Mod. Phys. 47 (1975) 849.

[143] A.A. Vladimirov, Method for computing renormalization group functions in dimensional renormalization scheme, Theor. Math. Phys. 43 (1980) 417.

[144] R. Coquereaux, Dimensional renormalization and comparison of renormalization schemes in Quantum Electrodynamics, Ann. Phys. 125 (1980) 401.

[145] S. Narison, Techniques of dimensional renormalization and applications to the two point functions of QCD and QED, Phys. Rep. 84 (1982) 263.

[146] P. Pascual, R. Tarrach, QCD: renormalization for the practitioner, Lect. Notes Phys. 194 (1984) 1.

[147] D.V. Shirkov, V.F. Kovalev, Bogolyubov renormalization group and symmetry of solution in mathematical physics, Phys. Rep. 352 (2001) 219, arXiv:hep-th/0001210.

[148] J.C. Collins, Renormalization: An Introduction to Renormalization, the Renormalization Group, and the Operator Product Expansion, Cambridge University Press, Cambridge, 1985, 380 pp.

[149] J.C. Collins, Renormalization: general theory, arXiv:hep-th/0602121.

[150] D.I. Kazakov, Radiative corrections, divergences, regularization, renormalization, renormalization group and all that in examples in Quantum Field Theory, arXiv:0901.2208 [hep-ph].

[151] G.M. Prosperi, M. Raciti, C. Simolo, On the running coupling constant in QCD, Prog. Part. Nucl. Phys. 58 (2007) 387, arXiv:hep-ph/0607209.

[152] A. Deur, S.J. Brodsky, G.F. de Teramond, The QCD running coupling, Prog. Part. Nucl. Phys. 90 (2016) 1, arXiv:1604.08082 [hep-ph].

[153] W.E. Caswell, Asymptotic behavior of non-abelian gauge theories to two-loop order, Phys. Rev. Lett. 33 (1974) 244.

[154] D.R.T. Jones, Two-loop diagrams in Yang-Mills theory, Nucl. Phys. B 75 (1974) 531.

[155] E. Egorian, O.V. Tarasov, Two-loop renormalization of the QCD in an arbitrary gauge, Theor. Math. Phys. 41 (1979) 863.

[156] O.V. Tarasov, A.A. Vladimirov, A. Zharkov, The Gell-Mann-low function of QCD in the three-loop approximation, Phys. Lett. B 93 (1980) 429.

[157] S.A. Larin, J.A.M. Vermaseren, The three-loop QCD β-function and anomalous dimensions, Phys. Lett. B 303 (1993) 334, arXiv:hep-ph/9302208.

[158] T. van Ritbergen, J.A.M. Vermaseren, S.A. Larin, The four-loop β-function in Quantum Chromodynamics, Phys. Lett. B 400 (1997) 379, arXiv:hep-ph/9701390.

[159] K.G. Chetyrkin, B.A. Kniehl, M. Steinhauser, Strong coupling constant with flavor thresholds at four loops in the $\overline{MS}$ scheme, Phys. Rev. Lett. 79 (1997) 2184, arXiv:hep-ph/9706430.

[160] P.A. Baikov, K.G. Chetyrkin, J.H. Kuhn, Five-loop running of the QCD coupling constant, arXiv:1606.08659 [hep-ph].

[161] W.A. Bardeen, A.J. Buras, D.W. Duke, T. Muta, Deep inelastic scattering beyond the leading order in asymptotically free gauge theories, Phys. Rev. D 18 (1978) 3998.

[162] A.N. Vasiliev, Y.M. Pismak, Y.R. Khonkonen, Simple method of calculating the critical indices in the $1/n$ expansion, Theor. Math. Phys. 46 (1981) 104.

[163] A.N. Vasiliev, Y.M. Pismak, Y.R. Khonkonen, $1/n$ expansion: calculation of the exponents η and ν in the order $1/n^2$ for arbitrary number of dimensions, Theor. Math. Phys. 47 (1981) 465.

[164] A.N. Vasiliev, Y.M. Pismak, Y.R. Khonkonen, $1/n$ expansion: calculation of the exponent η in the order $1/n^3$ by the conformal bootstrap method, Theor. Math. Phys. 50 (1982) 127.

[165] A.N. Vasiliev, M.Y. Nalimov, Analog of dimensional regularization for calculation of the renormalization group functions in the $1/n$ expansion for arbitrary dimension of space, Theor. Math. Phys. 55 (1983) 423.

[166] A.N. Vasiliev, M.Y. Nalimov, The CP^{n-1} model: calculation of anomalous dimensions and the mixing matrices in the order $1/n$, Theor. Math. Phys. 56 (1983) 643.

[167] A.N. Vasiliev, M.Y. Nalimov, Y.R. Khonkonen, $1/n$ expansion: calculation of anomalous dimensions and mixing matrices in the order $1/n$ for np matrix gauge invariant σ model, Theor. Math. Phys. 58 (1984) 111.

[168] J.A. Gracey, The QCD β function at $\mathcal{O}(1/n_f)$, Phys. Lett. B 373 (1996) 178, arXiv:hep-ph/9602214.

[169] M.A. Samuel, J.R. Ellis, M. Karliner, Comparison of the Pade approximation method to perturbative QCD calculations, Phys. Rev. Lett. 74 (1995) 4380, arXiv:hep-ph/9503411.

[170] S.J. Brodsky, J.R. Ellis, E. Gardi, M. Karliner, M.A. Samuel, Pade approximants, optimal renormalization scales, and momentum flow in Feynman diagrams, Phys. Rev. D 56 (1997) 6980, arXiv:hep-ph/9706467.

[171] J.R. Ellis, I. Jack, D.R.T. Jones, M. Karliner, M.A. Samuel, Asymptotic Pade approximant predictions: up to five loops in QCD and SQCD, Phys. Rev. D 57 (1998) 2665, arXiv:hep-ph/9710302.

[172] K. Symanzik, Infrared singularities and small distance behavior analysis, Commun. Math. Phys. 34 (1973) 7.

[173] T. Appelquist, J. Carazzone, Infrared singularities and massive fields, Phys. Rev. D 11 (1975) 2856.

[174] W. Bernreuther, Decoupling of heavy quarks in Quantum Chromodynamics, Ann. Phys. 151 (1983) 127.

[175] K.G. Chetyrkin, J.H. Kuhn, C. Sturm, QCD decoupling at four loops, Nucl. Phys. B 744 (2006) 121, arXiv:hep-ph/0512060.

[176] Y. Schroder, M. Steinhauser, Four-loop decoupling relations for the strong coupling, JHEP 601 (2006) 51, arXiv:hep-ph/0512058.

[177] B.A. Kniehl, A.V. Kotikov, A.I. Onishchenko, O.L. Veretin, Strong-coupling constant with flavor thresholds at five loops in the $\overline{MS}$ scheme, Phys. Rev. Lett. 97 (2006) 42001, arXiv:hep-ph/0607202.

[178] P.J. Redmond, Elimination of ghosts in propagators, Phys. Rev. 112 (1958) 1404.

[179] P.J. Redmond, J.L. Uretsky, Conjecture concerning the properties of nonrenormalizable field theories, Phys. Rev. Lett. 1 (1958) 147.

[180] N.N. Bogoliubov, A.A. Logunov, D.V. Shirkov, Dispersion relations and perturbation theory, Sov. Phys. JETP 37 (1960) 574.

[181] R. Brock, et al. [CTEQ Collaboration], Handbook of Perturbative QCD, Rev. Mod. Phys. 67 (1995) 157.

[182] S. Bethke, Experimental tests of asymptotic freedom, Prog. Part. Nucl. Phys. 58 (2007) 351, arXiv:hep-ex/0606035.

[183] S. Bethke, et al., Workshop on precision measurements of α_S, arXiv:1110.0016 [hep-ph].

[184] D. d'Enterria, P.Z. Skands, High-precision α_S measurements from LHC to FCC-ee, arXiv:1512.05194 [hep-ph].

[185] G. Rodrigo, A. Santamaria, QCD matching conditions at thresholds, Phys. Lett. B 313 (1993) 441, arXiv:hep-ph/9305305.

[186] K.G. Chetyrkin, B.A. Kniehl, M. Steinhauser, Decoupling relations to $\mathcal{O}(\alpha_s^3)$ and their connection to low-energy theorems, Nucl. Phys. B 510 (1998) 61, arXiv:hep-ph/9708255.

[187] W. Bernreuther, W. Wetzel, Decoupling of heavy quarks in the minimal subtraction scheme, Nucl. Phys. B 197 (1982) 228, 513 (1998) 758(E).

[188] S.A. Larin, T. van Ritbergen, J.A.M. Vermaseren, The large quark mass expansion of $\Gamma(Z^0 \to hadrons)$ and $\Gamma(\tau^- \to \nu_\tau + hadrons)$ in the order α_s^3, Nucl. Phys. B 438 (1995) 278, arXiv:hep-ph/9411260.

[189] W. Furmanski, R. Petronzio, Lepton-hadron processes beyond leading order in Quantum Chromodynamics, Z. Phys. C 11 (1982) 293.

[190] E. Gardi, M. Karliner, Relations between observables and the infrared fixed point in QCD, Nucl. Phys. B 529 (1998) 383, arXiv:hep-ph/9802218.

[191] E. Gardi, G. Grunberg, M. Karliner, Can the QCD running coupling have a causal analyticity structure?, JHEP 9807 (1998) 7, arXiv:hep-ph/9806462.

[192] T. Appelquist, A. Ratnaweera, J. Terning, L.C.R. Wijewardhana, The phase structure of an SU(N) gauge theory with n_f flavors, Phys. Rev. D 58 (1998) 105017, arXiv:hep-ph/9806472.

[193] B.A. Magradze, The gluon propagator in analytic perturbation theory, Conf. Proc. C 980518 (1999) 158, arXiv:hep-ph/9808247.

[194] A.V. Nesterenko, Investigation of a new analytic running coupling in QCD, Mod. Phys. Lett. A 15 (2000) 2401, arXiv:hep-ph/0102203.

[195] A.V. Nesterenko, Analytic invariant charge in QCD, Int. J. Mod. Phys. A 18 (2003) 5475, arXiv:hep-ph/0308288.

[196] Y.O. Belyakova, A.V. Nesterenko, A nonperturbative model for the strong running coupling within potential approach, Int. J. Mod. Phys. A 26 (2011) 981, arXiv:1011.1148 [hep-ph].

[197] T.L. Curtright, C.K. Zachos, Renormalization group functional equations, Phys. Rev. D 83 (2011) 65019, arXiv:1010.5174 [hep-th].

[198] H. Sonoda, Solving renormalization group equations with the Lambert W function, Phys. Rev. D 87 (2013) 85023, arXiv:1302.6069 [hep-th].

[199] R.M. Corless, G.H. Gonnet, D.E.G. Hare, D.J. Jeffrey, D.E. Knuth, On the Lambert W function, Adv. Comput. Math. 5 (1996) 329.

[200] D.J. Jeffrey, D.E.G. Hare, R.M. Corless, Unwinding the branches of the Lambert W function, Math. Sci. 21 (1996) 1.

[201] D. Veberic, Having fun with Lambert $W(x)$ function, arXiv:1003.1628 [cs.MS].

[202] S.R. Valluri, D.J. Jeffrey, R.M. Corless, Some applications of the Lambert W function to physics, Can. J. Phys. 78 (2000) 823.

[203] G. 't Hooft, M.J.G. Veltman, Regularization and renormalization of gauge fields, Nucl. Phys. B 44 (1972) 189.

[204] C.G. Bollini, J.J. Giambiagi, Dimensional renormalization: the number of dimensions as a regularizing parameter, Nuovo Cim. B 12 (1972) 20.

[205] G. 't Hooft, Dimensional regularization and the renormalization group, Nucl. Phys. B 61 (1973) 455.

[206] G. 't Hooft, An algorithm for the poles at dimension four in the dimensional regularization procedure, Nucl. Phys. B 62 (1973) 444.

[207] K.G. Chetyrkin, J.H. Kuhn, A. Kwiatkowski, QCD corrections to the e^+e^- cross-section and the Z boson decay rate: concepts and results, Phys. Rep. 277 (1996) 189.

[208] K.G. Chetyrkin, A.L. Kataev, F.V. Tkachov, New approach to evaluation of multiloop Feynman integrals: the Gegenbauer polynomial x space technique, Nucl. Phys. B 174 (1980) 345.

[209] K.G. Chetyrkin, F.V. Tkachov, Integration by parts: the algorithm to calculate β functions in 4 loops, Nucl. Phys. B 192 (1981) 159.

[210] T. Appelquist, H. Georgi, e^+e^- annihilation in gauge theories of strong interactions, Phys. Rev. D 8 (1973) 4000.

[211] A. Zee, Electron-positron annihilation in stagnant field theories, Phys. Rev. D 8 (1973) 4038.

[212] P.A. Baikov, K.G. Chetyrkin, J.H. Kuhn, $R(s)$ and hadronic τ-decays in order α_s^4: technical aspects, Nucl. Phys. B (Proc. Suppl.) 189 (2009) 49, arXiv:0906.2987 [hep-ph].

[213] P.A. Baikov, K.G. Chetyrkin, J.H. Kuhn, J. Rittinger, Vector correlator in massless QCD at order $\mathcal{O}(\alpha_s^4)$ and the QED β function at five loop, JHEP 1207 (2012) 17, arXiv:1206.1284 [hep-ph].

[214] K.G. Chetyrkin, A.L. Kataev, F.V. Tkachov, Higher-order corrections to σ_{tot} ($e^+e^- \to$ hadrons) in Quantum Chromodynamics, Phys. Lett. B 85 (1979) 277.

[215] M. Dine, J.R. Sapirstein, Higher-order Quantum Chromodynamic corrections in e^+e^- annihilation, Phys. Rev. Lett. 43 (1979) 668.

[216] W. Celmaster, R.J. Gonsalves, Analytic calculation of higher-order Quantum Chromodynamic corrections in e^+e^- annihilation, Phys. Rev. Lett. 44 (1980) 560.

[217] S.G. Gorishny, A.L. Kataev, S.A. Larin, The $\mathcal{O}(\alpha_s^3)$ corrections to σ_{tot} ($e^+e^- \to$ hadrons) and $\Gamma(\tau^- \to \nu_\tau +$ hadrons) in QCD, Phys. Lett. B 259 (1991) 144.

[218] L.R. Surguladze, M.A. Samuel, Total hadronic cross section in e^+e^- annihilation at the four-loop level of perturbative QCD, Phys. Rev. Lett. 66 (1991) 560, 2416(E).

[219] P.A. Baikov, K.G. Chetyrkin, J.H. Kuhn, Order α_s^4 QCD corrections to Z and τ decays, Phys. Rev. Lett. 101 (2008) 12002, arXiv:0801.1821 [hep-ph].

[220] P.A. Baikov, K.G. Chetyrkin, J.H. Kuhn, Adler function, Bjorken sum rule, and the Crewther relation to order α_s^4 in a general gauge theory, Phys. Rev. Lett. 104 (2010) 132004, arXiv:1001.3606 [hep-ph].

[221] P.A. Baikov, K.G. Chetyrkin, J.H. Kuhn, J. Rittinger, Adler function, sum rules and Crewther relation of order $\mathcal{O}(\alpha_s^4)$: the singlet case, Phys. Lett. B 714 (2012) 62, arXiv:1206.1288 [hep-ph].

[222] L.R. Surguladze, M.A. Samuel, Perturbative QCD calculations of total cross-sections and decay widths in hard inclusive processes, Rev. Mod. Phys. 68 (1996) 259, arXiv:hep-ph/9508351.

[223] P.A. Baikov, K.G. Chetyrkin, J.H. Kuhn, Massless propagators, $R(s)$ and multiloop QCD, Nucl. Part. Phys. Proc. 261 (2015) 3, arXiv:1501.06739 [hep-ph].

[224] K.G. Chetyrkin, J.H. Kuhn, Quartic mass corrections to R_{had}, Nucl. Phys. B 432 (1994) 337, arXiv:hep-ph/9406299.

[225] K.G. Chetyrkin, J.H. Kuhn, M. Steinhauser, Heavy quark vacuum polarization to three loops, Phys. Lett. B 371 (1996) 93, arXiv:hep-ph/9511430.

[226] K.G. Chetyrkin, J.H. Kuhn, M. Steinhauser, Three loop polarization function and $\mathcal{O}(\alpha_s^2)$ corrections to the production of heavy quarks, Nucl. Phys. B 482 (1996) 213, arXiv:hep-ph/9606230.

[227] A.H. Hoang, V. Mateu, S.M. Zebarjad, Heavy quark vacuum polarization function at $\mathcal{O}(\alpha_s^2)$ and $\mathcal{O}(\alpha_s^3)$, Nucl. Phys. B 813 (2009) 349, arXiv:0807.4173 [hep-ph].

[228] P.A. Baikov, K.G. Chetyrkin, J.H. Kuhn, Vacuum polarization in pQCD: first complete $\mathcal{O}(\alpha_s^4)$ result, Nucl. Phys. B (Proc. Suppl.) 135 (2004) 243.

[229] M.A. Shifman, A.I. Vainshtein, V.I. Zakharov, QCD and resonance physics. Theoretical foundations, Nucl. Phys. B 147 (1979) 385.

[230] M.A. Shifman, A.I. Vainshtein, V.I. Zakharov, QCD and resonance physics. Applications, Nucl. Phys. B 147 (1979) 448.

[231] M.A. Shifman, A.I. Vainshtein, V.I. Zakharov, QCD and resonance physics. The ρ-ω mixing, Nucl. Phys. B 147 (1979) 519.

[232] V.A. Novikov, M.A. Shifman, A.I. Vainshtein, V.I. Zakharov, Operator expansion in Quantum Chromodynamics beyond perturbation theory, Nucl. Phys. B 174 (1980) 378.

[233] V.A. Novikov, M.A. Shifman, A.I. Vainshtein, V.I. Zakharov, Calculations in external fields in Quantum Chromodynamics. Technical review, Fortsch. Phys. 32 (1984) 585.

[234] V.A. Novikov, M.A. Shifman, A.I. Vainshtein, M.B. Voloshin, V.I. Zakharov, Use and misuse of QCD sum rules, factorization and related topics, Nucl. Phys. B 237 (1984) 525.

[235] L.J. Reinders, H. Rubinstein, S. Yazaki, Hadron properties from QCD sum rules, Phys. Rep. 127 (1985) 1.

[236] S. Narison, QCD spectral sum rules, World Sci. Lect. Notes Phys. 26 (1989) 1.

[237] V.I. Zakharov, QCD perturbative expansions in large orders, Nucl. Phys. B 385 (1992) 452.

[238] E. de Rafael, An introduction to sum rules in QCD: course, arXiv:hep-ph/9802448.

[239] K.G. Chetyrkin, S. Narison, V.I. Zakharov, Short-distance tachyonic gluon mass and $1/Q^2$ corrections, Nucl. Phys. B 550 (1999) 353, arXiv:hep-ph/9811275.

[240] P. Colangelo, A. Khodjamirian, QCD sum rules, a modern perspective, arXiv:hep-ph/0010175.

[241] F.V. Gubarev, L. Stodolsky, V.I. Zakharov, On the significance of the vector potential squared, Phys. Rev. Lett. 86 (2001) 2220, arXiv:hep-ph/0010057.

[242] F.V. Gubarev, V.I. Zakharov, On the emerging phenomenology of $\langle A_{min}^2 \rangle$, Phys. Lett. B 501 (2001) 28, arXiv:hep-ph/0010096.

[243] H. Verschelde, K. Knecht, K. Van Acoleyen, M. Vanderkelen, The nonperturbative ground state of QCD and the local composite operator A_μ^2, Phys. Lett. B 516 (2001) 307, arXiv:hep-th/0105018.

[244] K.I. Kondo, Vacuum condensate of mass dimension 2 as the origin of mass gap and quark confinement, Phys. Lett. B 514 (2001) 335, arXiv:hep-th/0105299.

[245] S. Narison, V.I. Zakharov, Hints on the power corrections from current correlators in x space, Phys. Lett. B 522 (2001) 266, arXiv:hep-ph/0110141.

[246] S. Narison, The SVZ-expansion and beyond, Nucl. Phys. B (Proc. Suppl.) 164 (2007) 225, arXiv:hep-ph/0508259.

[247] O. Andreev, $1/q^2$ corrections and gauge/string duality, Phys. Rev. D 73 (2006) 107901, arXiv:hep-th/0603170.

[248] E.R. Arriola, W. Broniowski, Dimension-two gluon condensate from large-N_c Regge models, Phys. Rev. D 73 (2006) 97502, arXiv:hep-ph/0603263.

[249] S. Narison, V.I. Zakharov, Duality between QCD perturbative series and power corrections, Phys. Lett. B 679 (2009) 355, arXiv:0906.4312 [hep-ph].

[250] S. Narison, SVZ sum rules: $30 + 1$ years later, Nucl. Phys. B (Proc. Suppl.) 207 (2010) 315, arXiv:1010.1959 [hep-ph].

[251] D. Vercauteren, H. Verschelde, A two-component picture of the $\langle A_\mu^2 \rangle$ condensate with instantons, Phys. Lett. B 697 (2011) 70, arXiv:1101.5017 [hep-th].

[252] F. Jugeau, S. Narison, H. Ratsimbarison, SVZ$+1/q^2$-expansion versus some QCD holographic models, Phys. Lett. B 722 (2013) 111, arXiv:1302.6909 [hep-ph].

[253] P. Boucaud, M. Brinet, F. De Soto, V. Morenas, O. Pene, K. Petrov, J. Rodriguez-Quintero, Three-gluon running coupling from lattice QCD at $N_f = 2 + 1 + 1$: a consistency check of the OPE approach, JHEP 1404 (2014) 86, arXiv:1310.4087 [hep-ph].

[254] S. Randjbar-Daemi, J.A. Strathdee, Vacuum polarization and chiral lattice fermions, Nucl. Phys. B 461 (1996) 305, arXiv:hep-th/9510067.

[255] T. Blum, Lattice calculation of the lowest order hadronic contribution to the muon anomalous magnetic moment, Phys. Rev. Lett. 91 (2003) 52001, arXiv:hep-lat/0212018.

[256] C. Aubin, T. Blum, Calculating the hadronic vacuum polarization and leading hadronic contribution to the muon anomalous magnetic moment with improved staggered quarks, Phys. Rev. D 75 (2007) 114502, arXiv:hep-lat/0608011.

[257] C. Aubin, T. Blum, M. Golterman, S. Peris, Model-independent parametrization of the hadronic vacuum polarization and $g - 2$ for the muon on the lattice, Phys. Rev. D 86 (2012) 54509, arXiv:1205.3695 [hep-lat].

[258] C. Aubin, T. Blum, M. Golterman, S. Peris, Hadronic vacuum polarization with twisted boundary conditions, Phys. Rev. D 88 (2013) 74505, arXiv:1307.4701 [hep-lat].

[259] M. Golterman, K. Maltman, S. Peris, Tests of hadronic vacuum polarization fits for the muon anomalous magnetic moment, Phys. Rev. D 88 (2013) 114508, arXiv:1309.2153 [hep-lat].

[260] M. Gockeler, R. Horsley, W. Kurzinger, D. Pleiter, P.E.L. Rakow, G. Schierholz [QCDSF Collaboration], Vacuum polarization and hadronic contribution to muon $g - 2$ from lattice QCD, Nucl. Phys. B 688 (2004) 135, arXiv:hep-lat/0312032.

[261] E. Shintani, et al. [JLQCD and TWQCD Collaborations], T. Collaborations, Lattice study of the vacuum polarization function and determination of the strong coupling constant, Phys. Rev. D 79 (2009) 74510, arXiv:0807.0556 [hep-lat].

[262] E. Shintani, S. Aoki, H. Fukaya, S. Hashimoto, T. Kaneko, T. Onogi, N. Yamada, Strong coupling constant from vacuum polarization functions in three-flavor lattice QCD with dynamical overlap fermions, Phys. Rev. D 82 (2010) 74505, 89 (2014) 099903(E); arXiv:1002.0371 [hep-lat].

[263] P. Boyle, L. Del Debbio, E. Kerrane, J. Zanotti, Lattice determination of the hadronic contribution to the muon $g - 2$ using dynamical domain wall fermions, Phys. Rev. D 85 (2012) 74504, arXiv:1107.1497 [hep-lat].

[264] X. Feng, K. Jansen, M. Petschlies, D. Renner, Hadronic vacuum polarization contribution to $g - 2$ from the lattice, Nucl. Phys. B (Proc. Suppl.) 225 (2012) 269, arXiv:1112.4946 [hep-lat].

[265] X. Feng, S. Hashimoto, G. Hotzel, K. Jansen, M. Petschlies, D.B. Renner, Computing the hadronic vacuum polarization function by analytic continuation, Phys. Rev. D 88 (2013) 34505, arXiv:1305.5878 [hep-lat].

[266] X. Feng, S. Hashimoto, G. Hotzel, K. Jansen, M. Petschlies, D.B. Renner, Using analytic continuation for the hadronic vacuum polarization computation, arXiv:1311.0652 [hep-lat].

[267] M.D. Morte, B. Jager, A. Juttner, H. Wittig, The leading hadronic vacuum polarization on the lattice, AIP Conf. Proc. 1343 (2011) 337, arXiv:1011.5793 [hep-lat].

[268] M.D. Morte, B. Jager, H. Wittig, A. Juttner, Lattice determination of the anomalous magnetic moment of the muon, PoS LATTICE 161 (2011), arXiv:1111.2193 [hep-lat].

[269] M.D. Morte, B. Jager, A. Juttner, H. Wittig, Lattice calculations of the leading hadronic contribution to $(g-2)_\mu$, PoS LATTICE 175 (2012), arXiv:1211.1159 [hep-lat].

[270] M.D. Morte, B. Jager, A. Juttner, H. Wittig, Towards a precise lattice determination of the leading hadronic contribution to $(g-2)_\mu$, JHEP 1203 (2012) 55, arXiv:1112.2894 [hep-lat].

[271] A. Francis, B. Jager, H.B. Meyer, H. Wittig, A new representation of the Adler function for lattice QCD, Phys. Rev. D 88 (2013) 54502, arXiv:1306.2532 [hep-lat].

[272] H. Horch, G. Herdoiza, B. Jager, H. Wittig, M.D. Morte, A. Juttner, Computing the Adler function from the vacuum polarization function, PoS LATTICE 304 (2014), arXiv:1311.6975 [hep-lat].

[273] A. Francis, B. Jager, H.B. Meyer, H. Wittig, Adler function and hadronic vacuum polarization from lattice vector correlation functions in the time-momentum representation, PoS LATTICE 305 (2014), arXiv:1311.3285 [hep-lat].

[274] M.D. Morte, A. Francis, G. Herdoiza, H. Horch, B. Jager, A. Juttner, H. Meyer, H. Wittig, Study of the anomalous magnetic moment of the muon computed from the Adler function, PoS LATTICE 162 (2014), arXiv:1411.1206 [hep-lat].

[275] A. Francis, V. Gulpers, G. Herdoiza, H. Horch, B. Jager, H.B. Meyer, H. Wittig, Study of the hadronic contributions to the running of the QED coupling and the weak mixing angle, PoS LATTICE 110 (2015), arXiv:1511.04751 [hep-lat].

[276] M.D. Morte, G. Herdoiza, H. Horch, B. Jager, H. Meyer, H. Wittig, The hadronic vacuum polarization function with $\mathcal{O}(a)$-improved Wilson fermions—an update, PoS LATTICE 111 (2015), arXiv:1602.03976 [hep-lat].

[277] G. Bali, G. Endrodi, Hadronic vacuum polarization and muon $g-2$ from magnetic susceptibilities on the lattice, Phys. Rev. D 92 (2015) 54506, arXiv:1506.08638 [hep-lat].

[278] X.G. Wu, S.J. Brodsky, M. Mojaza, The renormalization scale-setting problem in QCD, Prog. Part. Nucl. Phys. 72 (2013) 44, arXiv:1302.0599 [hep-ph].

[279] X.G. Wu, Y. Ma, S.Q. Wang, H.B. Fu, H.H. Ma, S.J. Brodsky, M. Mojaza, Renormalization group invariance and optimal QCD renormalization scale setting, Rept. Prog. Phys. 78 (2015) 126201, arXiv:1405.3196 [hep-ph].

[280] A.L. Kataev, V.V. Starshenko, Estimates of the $\mathcal{O}(\alpha_s^4)$ corrections to $\sigma_{tot}(e^+e^- \to hadrons)$, $\Gamma(\tau \to \nu_\tau + hadrons)$ and deep inelastic scattering sum rules, arXiv:hep-ph/9405294.

[281] A.L. Kataev, V.V. Starshenko, Estimates of the higher order QCD corrections: theory and applications, Nucl. Phys. B (Proc. Suppl.) 39 (1995) 312, arXiv:hep-ph/9408395.

[282] P.A. Baikov, K.G. Chetyrkin, J.H. Kuhn, Towards order α_s^4 accuracy in τ decays, Phys. Rev. D 67 (2003) 74026, arXiv:hep-ph/0212299.

[283] G. Grunberg, Renormalization group improved perturbative QCD, Phys. Lett. B 95 (1980) 70.

[284] G. Grunberg, Renormalization-scheme-invariant QCD and QED: the method of effective charges, Phys. Rev. D 29 (1984) 2315.

[285] G. Grunberg, On some ambiguities in the method of effective charges, Phys. Rev. D 40 (1989) 680.

[286] G. Grunberg, Method of effective charges and BLM criterion, Phys. Rev. D 46 (1992) 2228.

[287] P.M. Stevenson, Resolution of the renormalization scheme ambiguity in perturbative QCD, Phys. Lett. B 100 (1981) 61.

[288] P.M. Stevenson, Optimized perturbation theory, Phys. Rev. D 23 (1981) 2916.

[289] W. Celmaster, P.M. Stevenson, Scale scheme ambiguities in the Brodsky–Lepage–Mackenzie procedure, Phys. Lett. B 125 (1983) 493.

[290] P.M. Stevenson, Optimization and the ultimate convergence of QCD perturbation theory, Nucl. Phys. B 231 (1984) 65.

[291] P.M. Stevenson, Explicit formula for the renormalization-scheme invariants of perturbation theory, Phys. Rev. D 33 (1986) 3130.

[292] A.C. Mattingly, P.M. Stevenson, QCD perturbation theory at low energies, Phys. Rev. Lett. 69 (1992) 1320, arXiv:hep-ph/9207228.

[293] A.C. Mattingly, P.M. Stevenson, Optimization of $R_{e^+e^-}$ and freezing of the QCD couplant at low energies, Phys. Rev. D 49 (1994) 437, arXiv:hep-ph/9307266.

[294] P.M. Stevenson, Optimization of QCD perturbation theory: results for $R_{e^+e^-}$ at fourth order, Nucl. Phys. B 868 (2013) 38, arXiv:1210.7001 [hep-ph].

[295] P.M. Stevenson, Fixed and unfixed points: infrared limits in optimized QCD perturbation theory, Nucl. Phys. B 875 (2013) 63, arXiv:1306.2371 [hep-ph].

[296] A. Dhar, Renormalization scheme-invariant perturbation theory, Phys. Lett. B 128 (1983) 407.

[297] A. Dhar, V. Gupta, New perturbative approach to renormalizable field theories, Phys. Rev. D 29 (1984) 2822.

[298] S.J. Brodsky, G.P. Lepage, P.B. Mackenzie, On the elimination of scale ambiguities in perturbative Quantum Chromodynamics, Phys. Rev. D 28 (1983) 228.

[299] G. Grunberg, A.L. Kataev, On some possible extensions of the Brodsky–Lepage–Mackenzie approach beyond the next-to-leading order, Phys. Lett. B 279 (1992) 352.

[300] J. Chyla, On the Brodsky–Lepage–Mackenzie scale-fixing procedure, its generalizations and the genuine higher order corrections, Phys. Lett. B 356 (1995) 341, arXiv:hep-ph/9505408.

[301] I.V. Anikin, B. Pire, L. Szymanowski, O.V. Teryaev, S. Wallon, On BLM scale fixing in exclusive processes, Eur. Phys. J. C 42 (2005) 163, arXiv:hep-ph/0411408.

[302] A.L. Kataev, S.V. Mikhailov, Generalization of the Brodsky–Lepage–Mackenzie optimization within the β-expansion and the principle of maximal conformality, Phys. Rev. D 91 (2015) 14007, arXiv:1408.0122 [hep-ph].

[303] A.L. Kataev, The generalized BLM approach to fix scale-dependence in QCD: the current status of investigations, J. Phys. Conf. Ser. 608 (2015) 12078, arXiv:1411.2257 [hep-ph].

[304] S.J. Brodsky, L. Di Giustino, Setting the renormalization scale in QCD: the principle of maximum conformality, Phys. Rev. D 86 (2012) 85026, arXiv:1107.0338 [hep-ph].

[305] S.J. Brodsky, X.G. Wu, Scale setting using the extended renormalization group and the principle of maximum conformality: the QCD coupling constant at four loops, Phys. Rev. D 85 (2012) 34038, 86 (2012) 079903(E); arXiv:1111.6175 [hep-ph].

[306] S.J. Brodsky, X.G. Wu, Eliminating the renormalization scale ambiguity for top-pair production using the principle of maximum conformality, Phys. Rev. Lett. 109 (2012) 42002, arXiv:1203.5312 [hep-ph].

[307] S.J. Brodsky, X.G. Wu, Application of the principle of maximum conformality to top-pair production, Phys. Rev. D 86 (2012) 14021, 87 (2013) 99902(E); arXiv:1204.1405 [hep-ph].

[308] S.J. Brodsky, X.G. Wu, Self-consistency requirements of the renormalization group for setting the renormalization scale, Phys. Rev. D 86 (2012) 54018, arXiv:1208.0700 [hep-ph].

[309] M. Mojaza, S.J. Brodsky, X.G. Wu, Systematic all-orders method to eliminate renormalization-scale and scheme ambiguities in perturbative QCD, Phys. Rev. Lett. 110 (2013) 192001, arXiv:1212.0049 [hep-ph].

[310] S.J. Brodsky, M. Mojaza, X.G. Wu, Systematic scale-setting to all orders: the principle of maximum conformality and commensurate scale relations, Phys. Rev. D 89 (2014) 14027, arXiv:1304.4631 [hep-ph].

[311] S.Q. Wang, X.G. Wu, S.J. Brodsky, Reanalysis of the higher order perturbative QCD corrections to hadronic Z decays using the principle of maximum conformality, Phys. Rev. D 90 (2014) 37503, arXiv:1406.1852 [hep-ph].

[312] A.E. Dorokhov, W. Broniowski, Vector and axial vector correlators in a nonlocal chiral quark model, Eur. Phys. J. C 32 (2003) 79, arXiv:hep-ph/0305037.

[313] A.E. Dorokhov, Adler function and hadronic contribution to the muon $g - 2$ in a nonlocal chiral quark model, Phys. Rev. D 70 (2004) 94011, arXiv:hep-ph/0405153.

[314] A.E. Dorokhov, Hadronic corrections to muon anomalous magnetic moment within the instanton liquid model, Acta Phys. Polon. B 36 (2005) 3751, arXiv:hep-ph/0510297.

[315] A.E. Dorokhov, Hadronic corrections to muon anomalous magnetic moment within the instanton liquid model, Nucl. Phys. A 790 (2007) 481.

[316] A.E. Dorokhov, Nonlocal condensates and current-current correlators within the instanton liquid model, arXiv:hep-ph/0601114.

[317] S. Peris, E. de Rafael, Low-energy QCD and ultraviolet renormalons, Nucl. Phys. B 500 (1997) 325, arXiv:hep-ph/9701418.

[318] C.J. Maxwell, Large order behavior of the QCD Adler D function in planar approximation, Phys. Lett. B 409 (1997) 382, arXiv:hep-ph/9706231.

[319] G. Cvetic, T. Lee, Bilocal expansion of Borel amplitude and hadronic tau decay width, Phys. Rev. D 64 (2001) 14030, arXiv:hep-ph/0101297.

[320] G. Cvetic, C. Dib, T. Lee, I. Schmidt, Resummation of the hadronic tau decay width with modified Borel transform method, Phys. Rev. D 64 (2001) 93016, arXiv:hep-ph/0106024.

[321] G. Cvetic, Techniques of evaluation of QCD low-energy physical quantities with running coupling with infrared fixed point, Phys. Rev. D 89 (2014) 36003, arXiv:1309.1696 [hep-ph].

[322] M. Beneke, M. Jamin, $\alpha(s)$ and the τ hadronic width: fixed-order, contour-improved and higher-order perturbation theory, JHEP 809 (2008) 44.

[323] A.L. Kataev, Is it possible to check urgently the 5-loop analytical results for the e^+e^--annihilation Adler function?, Phys. Lett. B 668 (2008) 350, arXiv:0808.3121 [hep-ph].

[324] A.L. Kataev, The analytical singlet α_s^4 QCD contributions into the e^+e^--annihilation Adler function and the generalized Crewther relations, JETP Lett. 94 (2011) 789, arXiv:1108.2898 [hep-ph].

[325] I. Caprini, J. Fischer, $\alpha(s)$ from τ decays: contour-improved versus fixed-order summation in a new QCD perturbation expansion, Eur. Phys. J. C 64 (2009) 35, arXiv:0906.5211 [hep-ph].

[326] I. Caprini, J. Fischer, Expansion functions in perturbative QCD and the determination of $\alpha_s(M_\tau^2)$, Phys. Rev. D 84 (2011) 54019, arXiv:1106.5336 [hep-ph].

[327] G. Abbas, B. Ananthanarayan, I. Caprini, Determination of $\alpha_s(M_\tau^2)$ from improved fixed order perturbation theory, Phys. Rev. D 85 (2012) 94018, arXiv:1202.2672 [hep-ph].

[328] G. Abbas, B. Ananthanarayan, I. Caprini, J. Fischer, Perturbative expansion of the QCD Adler function improved by renormalization-group summation and analytic continuation in the Borel plane, Phys. Rev. D 87 (2013) 14008, arXiv:1211.4316 [hep-ph].

[329] T. Goecke, C.S. Fischer, R. Williams, Leading-order calculation of hadronic contributions to the muon $g - 2$ using the Dyson–Schwinger approach, Phys. Lett. B 704 (2011) 211, arXiv:1107.2588 [hep-ph].

[330] B. Schrempp, F. Schrempp, QCD at low Q^2: a correspondence relation for moments of structure functions, Z. Phys. C 6 (1980) 7.

[331] K.A. Milton, I.L. Solovtsov, Analytic perturbation theory in QCD and Schwinger's connection between the beta function and the spectral density, Phys. Rev. D 55 (1997) 5295, arXiv:hep-ph/9611438.

[332] A.V. Nesterenko, J. Papavassiliou, The massive analytic invariant charge in QCD, Phys. Rev. D 71 (2005) 16009, arXiv:hep-ph/0410406.

[333] A.V. Nesterenko, J. Papavassiliou, The QCD analytic effective charge and its dependence on the pion mass, Int. J. Mod. Phys. A 20 (2005) 4622, arXiv:hep-ph/0409220.

[334] A.V. Nesterenko, J. Papavassiliou, Impact of the pion mass on nonpower expansion for QCD observables, Nucl. Phys. B (Proc. Suppl.) 164 (2007) 304, arXiv:hep-ph/0507320.

[335] A.V. Nesterenko, J. Papavassiliou, The QCD analytic running coupling and chiral symmetry breaking, Nucl. Phys. B (Proc. Suppl.) 152 (2005) 47, arXiv:hep-ph/0410072.

[336] A.V. Nesterenko, J. Papavassiliou, A novel integral representation for the Adler function, J. Phys. G 32 (2006) 1025, arXiv:hep-ph/0511215.

[337] A.V. Nesterenko, Dispersive approach to QCD and inclusive tau lepton hadronic decay, Phys. Rev. D 88 (2013) 56009, arXiv:1306.4970 [hep-ph].

[338] A.V. Nesterenko, Hadronic vacuum polarization function within dispersive approach to QCD, J. Phys. G 42 (2015) 85004, arXiv:1411.2554 [hep-ph].

[339] A.V. Nesterenko, On the low-energy behavior of the Adler function, Nucl. Phys. B (Proc. Suppl.) 186 (2009) 207, arXiv:0808.2043 [hep-ph].

[340] A.V. Nesterenko, Adler function in the analytic approach to QCD, SLAC eConf. C0706044 (2008) 25, arXiv:0710.5878 [hep-ph].

[341] A.I. Akhiezer, V.B. Berestetsky, Quantum Electrodynamics, Interscience, New York, 1965, 868 pp.

[342] S. Groote, J.G. Korner, A.A. Pivovarov, An interpolation of the vacuum polarization function for the evaluation of hadronic contributions to the muon anomalous magnetic moment, Eur. Phys. J. C 24 (2002) 393, arXiv:hep-ph/0111206.

[343] A.V. Nesterenko, The effects due to hadronization in the inclusive tau lepton decay, SLAC eConf. C1106064 (2011) 23, arXiv:1106.4006 [hep-ph].

[344] A.V. Nesterenko, Inclusive tau lepton decay: the effects due to hadronization, PoS ConfinementX 350 (2012), arXiv:1302.0518 [hep-ph].

[345] A.V. Nesterenko, Quark-antiquark potential in the analytic approach to QCD, Phys. Rev. D 62 (2000) 94028, arXiv:hep-ph/9912351.

[346] A.V. Nesterenko, New analytic running coupling in spacelike and timelike regions, Phys. Rev. D 64 (2001) 116009, arXiv:hep-ph/0102124.

[347] C. Contreras, G. Cvetic, O. Espinosa, H.E. Martinez, Simple analytic QCD model with perturbative QCD behavior at high momenta, Phys. Rev. D 82 (2010) 74005, arXiv:1006.5050 [hep-ph].

[348] C. Ayala, C. Contreras, G. Cvetic, Extended analytic QCD model with perturbative QCD behavior at high momenta, Phys. Rev. D 85 (2012) 114043, arXiv:1203.6897 [hep-ph].

[349] M.L. Perl, Lectures on electron-positron annihilation, Report SLAC-PUB-1614, 1975.

[350] A.V. Nesterenko, C. Simolo, QCDMAPT: program package for analytic approach to QCD, Comput. Phys. Commun. 181 (2010) 1769, arXiv:1001.0901 [hep-ph].

[351] A.V. Nesterenko, C. Simolo, QCDMAPT_F: Fortran version of QCDMAPT package, Comput. Phys. Commun. 182 (2011) 2303, arXiv:1107.1045 [hep-ph].

[352] A.V. Nesterenko, C. Simolo, in preparation.

[353] D.V. Shirkov, I.L. Solovtsov, Analytic QCD running coupling with finite IR behavior and universal $\alpha_S(0)$ value, arXiv:hep-ph/9604363.

[354] D.V. Shirkov, I.L. Solovtsov, Analytic model for the QCD running coupling with universal $\alpha_s(0)$ value, Phys. Rev. Lett. 79 (1997) 1209, arXiv:hep-ph/9704333.

[355] K.A. Milton, I.L. Solovtsov, Can the QCD effective charge be symmetrical in the Euclidean and the Minkowskian regions?, Phys. Rev. D 59 (1999) 107701, arXiv:hep-th/9812171.

[356] A.P. Bakulev, Global fractional analytic perturbation theory in QCD with selected applications, Phys. Part. Nucl. 40 (2009) 715, arXiv:0805.0829 [hep-ph].

[357] N.G. Stefanis, Taming Landau singularities in QCD perturbation theory: the analytic approach, Phys. Part. Nucl. 44 (2013) 494, arXiv:0902.4805 [hep-ph].

[358] G. Cvetic, A.V. Kotikov, Analogs of noninteger powers in general analytic QCD, J. Phys. G 39 (2012) 65005, arXiv:1106.4275 [hep-ph].

[359] P. Allendes, C. Ayala, G. Cvetic, Gluon propagator in fractional analytic perturbation theory, Phys. Rev. D 89 (2014) 54016, arXiv:1401.1192 [hep-ph].

[360] G. Cvetic, A.Y. Illarionov, B.A. Kniehl, A.V. Kotikov, Small-x behavior of the structure function F_2 and it slope $\partial \ln F_2/\partial \ln(1/x)$ for frozen and analytic strong-coupling constants, Phys. Lett. B 679 (2009) 350, arXiv:0906.1925 [hep-ph].

[361] A.V. Kotikov, Small x behavior of parton distributions. Analytical and frozen coupling constants. BFKL corrections, PoS (Baldin ISHEPP XXI) 033 (2013), arXiv:1212.3733 [hep-ph].

[362] A.V. Kotikov, B.G. Shaikhatdenov, Q^2-evolution of parton densities at small x values. Combined H1 and ZEUS F_2 data, Phys. Part. Nucl. 44 (2013) 543, arXiv:1212.4582 [hep-ph].

[363] A.V. Kotikov, B.G. Shaikhatdenov, Q^2 evolution of parton distributions at small values of x: effective scale for combined H1 and ZEUS data on the structure function F_2, Phys. Atom. Nucl. 78 (2015) 525, arXiv:1402.4349 [hep-ph].

[364] A.V. Kotikov, V.G. Krivokhizhin, B.G. Shaikhatdenov, Improved nonsinglet QCD analysis of fixed-target DIS data, J. Phys. G 42 (2015) 95004, arXiv:1411.1236 [hep-ph].

[365] R. Pasechnik, R. Enberg, G. Ingelman, QCD rescattering mechanism for diffractive deep inelastic scattering, Phys. Rev. D 82 (2010) 54036, arXiv:1005.3399 [hep-ph].

[366] O. Teryaev, Analyticity and higher twists, Nucl. Phys. B (Proc. Suppl.) 245 (2013) 195, arXiv:1309.1985 [hep-ph].

[367] A.V. Sidorov, O.P. Solovtsova, The QCD analysis of xF_3 structure function based on the analytic approach, Nonlin. Phenom. Complex Syst. 16 (2013) 397, arXiv:1312.3082 [hep-ph].

[368] A.V. Sidorov, O.P. Solovtsova, The QCD analysis of the combined set for the F_3 structure function data based on the analytic approach, Mod. Phys. Lett. A 29 (2014) 1450194, arXiv:1407.6858 [hep-ph].

[369] K.A. Milton, I.L. Solovtsov, O.P. Solovtsova, The Bjorken sum rule in the analytic approach to perturbative QCD, Phys. Lett. B 439 (1998) 421, arXiv:hep-ph/9809510.

[370] K.A. Milton, I.L. Solovtsov, O.P. Solovtsova, The Gross-Llewellyn Smith sum rule in the analytic approach to perturbative QCD, Phys. Rev. D 60 (1999) 16001, arXiv:hep-ph/9809513.

[371] R.S. Pasechnik, D.V. Shirkov, O.V. Teryaev, Bjorken sum rule and pQCD frontier on the move, Phys. Rev. D 78 (2008) 71902, arXiv:0808.0066 [hep-ph].

[372] R.S. Pasechnik, J. Soffer, O.V. Teryaev, Nucleon spin structure at low momentum transfers, Phys. Rev. D 82 (2010) 76007, arXiv:1009.3355 [hep-ph].

[373] V.L. Khandramai, R.S. Pasechnik, D.V. Shirkov, O.P. Solovtsova, O.V. Teryaev, Four-loop QCD analysis of the Bjorken sum rule vs data, Phys. Lett. B 706 (2012) 340, arXiv:1106.6352 [hep-ph].

[374] A.P. Bakulev, K. Passek-Kumericki, W. Schroers, N.G. Stefanis, Pion form-factor in QCD: from nonlocal condensates to NLO analytic perturbation theory, Phys. Rev. D 70 (2004) 33014, 70 (2004) 079906(E); arXiv:hep-ph/0405062.

[375] N.G. Stefanis, Pion form-factor analysis using NLO analytic perturbation theory, Nucl. Phys. B (Proc. Suppl.) 152 (2006) 245, arXiv:hep-ph/0410245.

[376] A.P. Bakulev, A.V. Pimikov, N.G. Stefanis, QCD sum rules with nonlocal condensates and the spacelike pion form factor, Phys. Rev. D 79 (2009) 93010, arXiv:0904.2304 [hep-ph].

[377] A.P. Bakulev, A.V. Pimikov, N.G. Stefanis, Pion form factor in the QCD sum-rule approach with nonlocal condensates, Acta Phys. Polon. Supp. B 2 (2009) 289, arXiv:0905.2522 [hep-ph].

[378] M. Baldicchi, G.M. Prosperi, Infrared behavior of the running coupling constant and bound states in QCD, Phys. Rev. D 66 (2002) 74008, arXiv:hep-ph/0202172.

[379] M. Baldicchi, G.M. Prosperi, Running coupling constant and masses in QCD, the meson spectrum, AIP Conf. Proc. 756 (2005) 152, arXiv:hep-ph/0412359.

[380] M. Baldicchi, G.M. Prosperi, C. Simolo, Extracting infrared QCD coupling from meson spectrum, AIP Conf. Proc. 892 (2007) 340, arXiv:hep-ph/0611087.

[381] M. Baldicchi, A.V. Nesterenko, G.M. Prosperi, D.V. Shirkov, C. Simolo, Bound state approach to the QCD coupling at low energy scales, Phys. Rev. Lett. 99 (2007) 242001, arXiv:0705.0329 [hep-ph].

[382] M. Baldicchi, A.V. Nesterenko, G.M. Prosperi, C. Simolo, QCD coupling below 1 GeV from quarkonium spectrum, Phys. Rev. D 77 (2008) 34013, arXiv:0705.1695 [hep-ph].

[383] C. Ayala, G. Cvetic, Calculation of binding energies and masses of quarkonia in analytic QCD models, Phys. Rev. D 87 (2013) 54008, arXiv:1210.6117 [hep-ph].

[384] A.V. Nesterenko, Analytic invariant charge and the lattice static quark-antiquark potential, Int. J. Mod. Phys. A 19 (2004) 3471, arXiv:hep-ph/0305091.

[385] A.V. Nesterenko, Quark confinement in the analytic approach to QCD, in: W. Lucha, K.M. Maung (Eds.), Proceedings of 4th International Conference on Quark Confinement and the Hadron Spectrum, Vienna, Austria, 2000, World Scientific Publishing Co., 2002, p. 255, hep-ph/0010257.

[386] A.V. Nesterenko, Dispersive approach to QCD: τ lepton hadronic decay in vector and axial-vector channels, Nucl. Part. Phys. Proc. 258 (2015) 177, arXiv:1409.0687 [hep-ph].

[387] A.V. Nesterenko, Inclusive τ lepton hadronic decay in vector and axial-vector channels within dispersive approach to QCD, AIP Conf. Proc. 1701 (2016) 40016, arXiv:1508.03705 [hep-ph].

[388] A.V. Nesterenko, Inclusive tau lepton hadronic decay in the framework of dispersive approach to QCD, in: A.I. Studenikin (Ed.), Proceedings of 16th Lomonosov Conference on Elementary Particle Physics, Moscow, Russia, 2013, World Scientific Publishing Co., 2015, p. 349, arXiv:1401.0620 [hep-ph].

[389] A.V. Nesterenko, Hadronic effects in low-energy QCD: inclusive tau lepton decay, Nucl. Phys. B (Proc. Suppl.) 234 (2013) 199, arXiv:1209.0164 [hep-ph].

[390] A.V. Nesterenko, Hadronization effects in inclusive tau decay, in: A.I. Studenikin (Ed.), Proceedings of 15th Lomonosov Conference on Elementary Particle Physics, Moscow, Russia, 2011, World Scientific Publishing Co., 2013, p. 363, arXiv:1110.3415 [hep-ph].

[391] K.A. Milton, I.L. Solovtsov, O.P. Solovtsova, Analytic perturbation theory and inclusive tau decay, Phys. Lett. B 415 (1997) 104, arXiv:hep-ph/9706409.

[392] K.A. Milton, I.L. Solovtsov, O.P. Solovtsova, V.I. Yasnov, Renormalization scheme and higher loop stability in hadronic tau decay within analytic perturbation theory, Eur. Phys. J. C 14 (2000) 495, arXiv:hep-ph/0003030.

[393] G. Cvetic, C. Villavicencio, Operator product expansion with analytic QCD in tau decay physics, Phys. Rev. D 86 (2012) 116001, arXiv:1209.2953 [hep-ph].

[394] A.C. Aguilar, A.V. Nesterenko, J. Papavassiliou, Infrared enhanced analytic coupling and chiral symmetry breaking in QCD, J. Phys. G 31 (2005) 997, arXiv:hep-ph/0504195.

[395] A.C. Aguilar, A.V. Nesterenko, J. Papavassiliou, Quark gap equation within the analytic approach to QCD, Nucl. Phys. B (Proc. Suppl.) 164 (2007) 300, arXiv:hep-ph/0510117.

[396] N. Christiansen, M. Haas, J.M. Pawlowski, N. Strodthoff, Transport coefficients in Yang-Mills theory and QCD, Phys. Rev. Lett. 115 (2015) 112002, arXiv:1411.7986 [hep-ph].

[397] N. Mueller, J.M. Pawlowski, Magnetic catalysis and inverse magnetic catalysis in QCD, Phys. Rev. D 91 (2015) 116010, arXiv:1502.08011 [hep-ph].

[398] R. Lang, N. Kaiser, W. Weise, Shear viscosities from Kubo formalism in a large-N_c Nambu–Jona–Lasinio model, Eur. Phys. J. A 51 (2015) 127, arXiv:1506.02459 [hep-ph].

[399] A.I. Alekseev, B.A. Arbuzov, Analyticity and minimality of nonperturbative contributions in perturbative region for α_s, Mod. Phys. Lett. A 13 (1998) 1747, arXiv:hep-ph/9704228.

[400] A.I. Alekseev, B.A. Arbuzov, An invariant charge model for all $q^2 > 0$ in QCD and gluon condensate, Mod. Phys. Lett. A 20 (2005) 103, arXiv:hep-ph/0411339.

[401] G. Cvetic, C. Valenzuela, Various versions of analytic QCD and skeleton-motivated evaluation of observables, Phys. Rev. D 74 (2006) 114030, 84 (2011) 019902(E); arXiv:hep-ph/0608256.

[402] G. Cvetic, R. Kogerler, C. Valenzuela, Reconciling the analytic QCD with the ITEP operator product expansion philosophy, Phys. Rev. D 82 (2010) 114004, arXiv:1006.4199 [hep-ph].

[403] G. Cvetic, R. Kogerler, C. Valenzuela, Analytic QCD coupling with no power terms in UV regime, J. Phys. G 37 (2010) 75001, arXiv:0912.2466 [hep-ph].

[404] G. Cvetic, R. Kogerler, Applying generalized Pade approximants in analytic QCD models, Phys. Rev. D 84 (2011) 56005, arXiv:1107.2902 [hep-ph].

[405] K.A. Milton, I.L. Solovtsov, O.P. Solovtsova, The Adler function for light quarks in analytic perturbation theory, Phys. Rev. D 64 (2001) 16005, arXiv:hep-ph/0102254.

[406] K.A. Milton, I.L. Solovtsov, O.P. Solovtsova, An analytic method of describing R-related quantities in QCD, Mod. Phys. Lett. A 21 (2006) 1355, arXiv:hep-ph/0512209.

[407] G. Cvetic, C. Valenzuela, I. Schmidt, A modification of minimal analytic QCD at low energies, Nucl. Phys. B (Proc. Suppl.) 164 (2007) 308, arXiv:hep-ph/0508101.

[408] I.L. Solovtsov, D.V. Shirkov, Analytic approach in Quantum Chromodynamics, Theor. Math. Phys. 120 (1999) 1220, arXiv:hep-ph/9909305.

[409] D.V. Shirkov, Analytic perturbation theory in analyzing some QCD observables, Eur. Phys. J. C 22 (2001) 331, arXiv:hep-ph/0107282.

[410] D.V. Shirkov, I.L. Solovtsov, Ten years of the analytic perturbation theory in QCD, Theor. Math. Phys. 150 (2007) 132, arXiv:hep-ph/0611229.

[411] G. Cvetic, C. Valenzuela, Analytic QCD: a short review, Braz. J. Phys. 38 (2008) 371, arXiv:0804.0872 [hep-ph].

[412] F. Schrempp, Tracking QCD instantons, J. Phys. G 28 (2002) 915, arXiv:hep-ph/0109032.

[413] D. Klammer, F. Schrempp, QCD-instantons and conformal space-time inversion symmetry, JHEP 806 (2008) 98, arXiv:0804.4573 [hep-ph].

[414] D.A. Smith, M.J. Teper [UKQCD Collaboration], Topological structure of the SU(3) vacuum, Phys. Rev. D 58 (1998) 14505, arXiv:hep-lat/9801008.

[415] A.V. Nesterenko, I.L. Solovtsov, New analytic running coupling in QCD: higher loop levels, Mod. Phys. Lett. A 16 (2001) 2517, arXiv:hep-ph/0111067.

[416] D.M. Howe, C.J. Maxwell, All-orders infrared freezing of $R_{e^+e^-}$ in perturbative QCD, Phys. Lett. B 541 (2002) 129, arXiv:hep-ph/0204036.

[417] D.M. Howe, C.J. Maxwell, All-orders infrared freezing of observables in perturbative QCD, Phys. Rev. D 70 (2004) 14002, arXiv:hep-ph/0303163.

[418] P.M. Brooks, C.J. Maxwell, Infrared freezing of Euclidean QCD observables, Phys. Rev. D 74 (2006) 65012, arXiv:hep-ph/0604267.

[419] I. Caprini, J. Fischer, On the infrared freezing of perturbative QCD in the Minkowskian region, Phys. Rev. D 71 (2005) 94017, arXiv:hep-ph/0505016.

[420] I. Caprini, J. Fischer, Comment on infrared freezing of Euclidean QCD observables, Phys. Rev. D 76 (2007) 18501, arXiv:0706.2914 [hep-ph].

[421] N.V. Krasnikov, A.A. Pivovarov, Running coupling at small momenta, renormalization schemes and renormalons, Phys. Atom. Nucl. 64 (2001) 1500, arXiv:hep-ph/9510207.

[422] T. Kugo, I. Ojima, Local covariant operator formalism of non-abelian gauge theories and quark confinement problem, Prog. Theor. Phys. Suppl. 66 (1979) 1.

[423] T. Kugo, The universal renormalization factors Z_1/Z_3 and color confinement condition in non-abelian gauge theory, arXiv:hep-th/9511033.

[424] H. Hata, Restoration of the local gauge symmetry and color confinement in non-abelian gauge theories, Prog. Theor. Phys. 67 (1982) 1607.

[425] H. Hata, Restoration of the local gauge symmetry and color confinement in non-abelian gauge theories II, Prog. Theor. Phys. 69 (1983) 1524.

[426] K.I. Kondo, Kugo-Ojima color confinement criterion and Gribov-Zwanziger horizon condition, Phys. Lett. B 678 (2009) 322, arXiv:0904.4897 [hep-th].

[427] A.C. Aguilar, D. Binosi, J. Papavassiliou, Indirect determination of the Kugo-Ojima function from lattice data, JHEP 0911 (2009) 66, arXiv:0907.0153 [hep-ph].

[428] P. Watson, R. Alkofer, Verifying the Kugo-Ojima confinement criterion in Landau gauge QCD, Phys. Rev. Lett. 86 (2001) 5239, arXiv:hep-ph/0102332.

[429] R. Alkofer, C.S. Fischer, L. von Smekal, Kugo-Ojima confinement and QCD Green's functions in covariant gauges, Prog. Part. Nucl. Phys. 50 (2003) 317, arXiv:nucl-th/0301048.

[430] D.V. Shirkov, The π^2-terms in the s-channel QCD observables, arXiv:hep-ph/0009106.

[431] A.V. Nesterenko, S.A. Popov, R-ratio of electron-positron annihilation into hadrons: higher-order pi2-terms, Nucl. Part. Phys. Proc. (to be published), arXiv:1608.04205 [hep-ph].

[432] D.S. Kourashev, B.A. Magradze, Explicit expressions for Euclidean and Minkowskian QCD observables in analytic perturbation theory, Theor. Math. Phys. 135 (2003) 531, arXiv:hep-ph/0104142.

[433] B.A. Magradze, Practical techniques of analytic perturbation theory of QCD, arXiv:hep-ph/0305020.

[434] D.V. Shirkov, A.V. Zayakin, Analytic perturbation theory for practitioners and upsilon decay, Phys. Atom. Nucl. 70 (2007) 775, arXiv:hep-ph/0512325.

[435] A.P. Bakulev, V.L. Khandramai, FAPT: a Mathematica package for calculations in QCD fractional analytic perturbation theory, Comput. Phys. Commun. 184 (2013) 183, arXiv:1204.2679 [hep-ph].

[436] C. Ayala, G. Cvetic, anQCD: a Mathematica package for calculations in general analytic QCD models, Comput. Phys. Commun. 190 (2015) 182, arXiv:1408.6868 [hep-ph].

[437] C. Ayala, G. Cvetic, Mathematica and Fortran programs for various analytic QCD couplings, J. Phys. Conf. Ser. 608 (2015) 12064, arXiv:1411.1581 [hep-ph].

[438] C. Ayala, G. Cvetic, anQCD: Fortran programs for couplings at complex momenta in various analytic QCD models, Comput. Phys. Commun. 199 (2016) 114, arXiv:1506.07201 [hep-ph].

Index

Printed in the United States
By Bookmasters